"The Human Psyche" is an in-depth exploration of dualist-interactionism, a concept Sir John Eccles developed with Sir Karl Popper, in the context of a wide variety of brain activities relating to self-conciousness. Opening with a critical discussion of materialist hypotheses on this relationship, it aims to demonstrate the great explanatory power of dualist-interactionism in contrast to the poverty and inadequacy of materialist theories of the mind. This strong version of dualist interactionism is demonstrated to be coherent and in general accord with present scientific knowledge.

The closing sections of the book show that dualist-interactionism offers valuable insights into the higher levels of human experience that cannot be accommodated to materialist theories of the mind – for example: the whole range of values; freedom of the will and moral responsibility; the uniqueness of the human person; the quest of meaning and for hope in the context of the inevitable end in death of this, our life on earth.

This unique work, reaching far beyond the neurosciences, will be of profound interest to all those concerned with the workings and the products of the human mind.

ALLE ZEIT WACH 1842

John C. Eccles

The Human Psyche

The GIFFORD Lectures
University of Edinburgh 1978–1979

With 76 Figures

Springer International 1980

Sir John Eccles
CH-6611 Contra (Locarno) TI, Switzerland
Max-Planck-Institut für biophysikalische Chemie
(Karl-Friedrich-Bonnhoeffer-Institut) D-3400 Göttingen, West Germany

ISBN-13: 978-3-642-49254-9 e-ISBN-13: 978-3-642-49252-5
DOI: 10.1007/978-3-642-49252-5

Library of Congress Cataloging in Publication Data. Eccles, John Carew, Sir. The human psyche. (Gifford lectures; 1979) Bibliography: p. Includes index. 1. Brain. 2. Mind and body. 3. Neuropsychology. 4. Anthroposophy. I. Title. II. Series. QP376.E266 128'.2 80-16980

Softcover reprint of the hardcover 1st edition 1980

2120/3130-543210

From the Ode to PSYCHE by John Keats

Yes, I will be thy priest, and build a fane
  In some untrodden region of my mind,
Where branched thoughts, new grown with pleasant pain,
  Instead of pines shall murmur in the wind:

And in the midst of this wide quietness
  A rosy sanctuary will I dress
With the wreath'd trellis of a working brain,
  With buds, and bells, and stars without a name,
With all the gardener Fancy e'er could feign,
  Who breeding flowers, will never breed the same:
And there shall be for thee all soft delight
  That shadowy thought can win,
A bright torch, and a casement ope at night,
  To let the warm Love in!

For Helena

my Muse

Preface

In February and March 1978 I delivered my first series of Gifford Lectures in the University of Edinburgh. These lectures have been published under the title *The Human Mystery*. The second series of ten lectures were delivered from April 18 to May 4 1979 under the title *The Human Psyche*. As with the first series, the printed text is actually the manuscript prepared for those lectures, not some later compilation. The lectures were delivered informally, but based strictly on this manuscript. It is hoped that the printed text will convey the dramatic character of a lecture presentation. This book must not be regarded as a definitive text in neuroscience, psychology and philosophy, but rather as a series of 'adventures of ideas', to revive a Whiteheadean title.

The brain–mind problem has been the theme of three recent books: *The Self and Its Brain; The Human Mystery* (in its latter part); and now *The Human Psyche*. In this book there is critical discussion in the first lecture of the materialist hypotheses of the relationship of the self-conscious mind to the brain. In the subsequent lectures the strong dualist-interactionism developed in *The Self and Its Brain* is explored in depth in relation to a wide variety of phenomena relating to self-consciousness. The aim has been to demonstrate the great explanatory power of dualist-interactionism in contrast to the poverty and inadequacy of all varieties of the materialist theories of the mind. It is not claimed that this strong version of dualist-interactionism is thereby authenticated; but at least it is demonstrated to be coherent and to be in general accord with our present scientific knowledge, which of course is not well developed in relationship to the phenomena of conscious experience. Moreover the last four lectures demonstrate that dualist-interactionism offers valuable insights into the higher levels of human experience that cannot be accommodated to materialist theories of the mind. One can list for example: the whole range of values; freedom of the will and moral responsibility; the uniqueness of the human person; the quest for meaning and for hope in the context of the inevitable end in death of this, our life on earth.

The exploration of the brain–mind relationship is based on the scientific study of the brain, both its structure and its physiological performance. So far as possible the anatomical study has been based on the primate brain with particular emphasis on the recent demonstration of its organization in biological units or modules. Moreover on the functional side there are now remarkable studies on the human brain in perception, attention, intention, emotion, levels of consciousness, learning and memory. These studies have been carried out on normal subjects as well as on those with cerebral lesions, most notably commissurotomy. Thus the human psyche has been considered in a scientific manner in what we may call psychoneurology. In the later chapters the theme develops more philosophical overtones in relation to such topics as creativity, altruism, pseudaltruism, aggression and values. Finally considerations of the purpose and meaning of self-conscious life lead to the climax in the unifying concepts of the human psyche and its status in relation to God. Thus the whole lecture series can be considered in the words by Lord Gifford: '... to promote and diffuse the study of Natural Theology in the widest sense of that term – in other words, the knowledge of God'.

The last century of philosophical and religious catharsis has stripped mankind of the religious concept of the soul and the deeper spiritual meaning of personal life. The outcome of this predicament tends to be hedonism or the nihilism of despair. In these two series of Gifford Lectures there has been a re-examination and a criticism of the materialist and anti-religious philosophies that form so much of the academic establishment in this age of disillusionment, dominated as it is by science. It is not the purpose of these lectures to discredit science – on the contrary! But it is the purpose to attack and discredit scientism, which may be defined as the effort to extend science beyond its proper sphere of operation. The arrogant claim of scientism is that science will soon in principle be able to provide a complete explanation of all our experiences, not only of the material world, but also of the world of our inner sensing in respect of values, of ideals, and of our most subtle and noble feelings of beauty, friendship and love as expressed and enshrined in the great masterpieces of human creativity.

The title of the book *The Human Psyche* was chosen because the theme progresses from a study of the brain–mind relationship as revealed in experimental neurology to the concept of the soul or psyche as originally expressed in the Platonic dialogues – in particular *The Phaedo* – and as further developed in Christianity. The Greeks symbolically represented Psyche as the youngest of the goddesses. As such Psyche has inspired many artists. Most notable is *The Story of Psyche* painted by Raphael and Giulio Romano on the ceiling of the Loggia of the Villa Farnesina in Rome. Psyche has been the subject of one of the great odes by John Keats. The motive of this book is quoted from the last stanza of the Ode to Psyche. It reveals his subtle poetic thought on the brain–mind problem.

It is my hope that the philosophy expressed in these Gifford Lectures will help to restore to human persons a belief in their spiritual nature superimposed on their material body and brain. With that restoration there will come a religious illumination giving hope and meaning to their ineffable existence as conscious selves. It has not been my task as a Gifford Lecturer to consider the details of religious dogmas and beliefs. But I have endeavoured to show that the philosophy of dualist-interactionism leads to a belief in the primacy of the spiritual nature of man, which in its turn leads 'on to God' as Max Planck so emphatically says (Quotation on p. 247).

JOHN C. ECCLES

Acknowledgements

I wish to express my thanks to the Gifford Lectureship Committee for the invitation to deliver the Gifford Lectures at the University of Edinburgh during the academic year 1978–1979. I would like to thank especially Professor Tom Torrance who initiated the invitation, our host, Professor J. McIntyre, the Acting Principal of the University of Edinburgh and Miss Jean Ewan, Secretary of the Gifford Lectureship Committee, for so efficiently and kindly making all arrangements before and during the lectures. I need not name the many friends who made the stay in Edinburgh so memorable for my wife and myself. I would also like to express my gratitude to my audience of up to 250 who provided an enthusiastic atmosphere, and who stayed the intensive course of 10 lectures in $2^1/_2$ weeks.

Acknowledgements are given in the text for the many excellent illustrations that form an important part of such a wide ranging scientific presentation. My special thanks are due to Professor Rolf Hassler and Dr. Manfred Klee for so kindly arranging for the preparation of the illustrations which was carried out so well by Hedwig Thomas and Jan Jaeger.

Again the staff of Springer-Verlag has given very personal service in the publishing of this book. Dr. Heinz Götze has been a very good friend throughout the whole publishing operation, and I wish particularly to thank Dr. Thomas Thiekötter for his dedication and efficiency.

Contents

Lecture 1

Consciousness, Self-consciousness and the Brain–Mind Problem

Résumé

This lecture is partly in overlapping relationship with the last lectures of my first Gifford series. The usage of the word *self-consciousness* in contrast to *consciousness* is considered along with the problem of the development of self-consciousness out of animal consciousness.

The experienced *unity of self-consciousness* is discussed in relation to the two cerebral hemispheres and in particular to the very recent results obtained by Sperry and his associates using the most subtle and sensitive methods for detection of self-conscious responses of the right hemispheres of commissurotomized patients. This 'isolated' hemisphere seems to have a limited self-consciousness in its recognition of faces and objects with appropriate emotional reactions. The various interpretations of these extraordinary results are fully discussed.

The second part of the lecture is devoted to a formulation of the several hypotheses relating to the brain–mind problem and to a critical discussion of the so-called *materialist theories of the mind.* Two reasons are given for rejecting the claim that these materialist theories are in accord with natural law as it now is. In addition it is claimed that all materialist theories are reducible to determinism which entails a negation of rational discussion. This theme is further developed in Lecture 10. There is formulation and discussion of the alternative hypothesis of *dualist-interactionism* that was developed in the Lecture 10 of the last series. Finally there is a critical evaluation of the diverse hypotheses formulated by neuroscientists in relation to the brain–mind problem.

1.1 Introduction

Last year in my Gifford Lectures (now published as *The Human Mystery*, Eccles, 1979a) I traced the line of contingency from the Big Bang to our existence here and now. From that wonderful series of happenings in all their mysterious sequences, I derived much of relevance to the over-riding theme of Natural Theology. The last three lectures dealt in some detail with the structure of the human brain, with conscious perception, with cognitive memory and finally with the brain–mind problem. This series of lectures is planned to continue logically from the first series by considering the wide range of manifestations of consciousness and their relation to the activities in the neuronal machinery of the brain. Hence it is advisable to overlap a little with the last series of lectures in order to establish a continuity of theme between the two series.

Following the procedure adopted by Popper and Eccles (1977) it is proposed to use the term *self-conscious mind* for the highest mental experiences. It implies knowing that one knows, which is of course initially a subjective or introspective criterion. However by linguistic communication it can be authenticated that other human beings share in this experience of self-knowing. One has only to listen to ordinary conversation, which is largely devoted to recounting the conscious experiences of the speakers. At a lower level there can be consciousness or awareness as indicated by intelligent learned behaviour and by emotional reactions. We can speak of an animal as conscious when it is capable of assessing the complexities of its present situation in the light of past experience and so is able to arrive at an appropriate course of action that is more than a stereotyped instinctive response. In this way it can exhibit an original behaviour pattern which can be learnt, and also which includes a wealth of emotional reactions. Reference should be made to the excellent accounts by Wilson (1975), Thorpe (1974), and Griffin (1976).

You may well ask: when does self-consciousness develop out of such a consciousness? A test for self can be identification in a mirror. Gallup (1977) has found that a chimpanzee can learn with difficulty to recognize itself in a mirror as shown by its use of the mirror image to remove a coloured mark on its face. Monkeys never learn in this way and there are no reported examples with other mammals. So it would seem that anthropoid apes have some primitive knowledge of self, but, as discussed in the last Gifford Lectures (Eccles, 1979a, Chap. 6), a fully developed recognition of self can only be demonstrated in the archeological records of ceremonial burial by Neanderthal man some 80,000 years ago. It can be anticipated that further discoveries will place this critical time much earlier, particularly in view of the evidence from anthropoid apes.

But the above question has also to be asked for the developing human being. The investigations of Amsterdam (1972) lead him to give 18 months for the transition from the conscious baby to the self-conscious child.

Awareness is commonly used as a clinical term to signify that a patient is able to respond to verbal commands and to visual and cutaneous stimuli. One can also use the term self-awareness instead of self-consciousness, but I prefer self-consciousness because it relates directly to the self-conscious mind. However it must not be concluded that the use of this substantive term implies the recognition of mind as a substance in the Cartesian manner. Rather we can refer to mind as an entity. It comprises subjective experiences of all kinds and thus is identical with World 2 as defined by Popper (cf. Fig. 7-1).

The present series of lectures on 'The Human Psyche' is necessarily based on the full range of experiences that relate to the conscious self, though concepts of subconsciousness are also explored. There is still the tendency for materialist philosophers to denigrate the term 'mind' to 'minding' and to reject the substantive use of mind, as was done by Ryle (1949) in his influential book *The Concept of Mind.* It appeared at that time that the word mind had been finally exorcised, it being sufficient to characterize our experiences of human beings in descriptive terms of their actions and of their verbal behaviour. However Ryle was soon answered in a remarkable book by Beloff (1962) – *The Existence of Mind.*

Though the disciples of Ryle still actively promulgate reductionism in the various materialist theories of the mind (Smart, 1963; Armstrong, 1968; Blakemore, 1977), there has been from the time of Beloff's book a remarkable reaction in a literature that ranges from a sober re-evaluation of the mind–brain relationships studied in various clinical conditions, for example in the writings of Penfield (1975), Sperry (1976; 1977), Zangwill (1976) and their associates to a veritable flowering of poetic imagery by Jaynes (1976), who begins his extraordinary book *On the origin of consciousness in the breakdown of the bicameral mind* by a pæan of praise:

> O, what a world of unseen visions and heard silences, this insubstantial country of the mind! What ineffable essences, these touchless rememberings and unshowable reveries! And the privacy of it all! A secret theater of speechless monologue and prevenient counsel, an invisible mansion of all moods, musings, and mysteries, an infinite resort of disappointments and discoveries. A whole kingdom where each of us reigns reclusively alone, questioning what we will, commanding what we can. A hidden hermitage where we may study out the troubled book of what we have done and yet may do. An introcosm that is more myself than anything I can find in a mirror. This consciousness that is myself of selves, that is everything, and yet nothing at all – what is it?
> And where did it come from?
> And why?

We can contrast this rhapsody with the testing procedures of experimental neuropsychology with its coloured patches or gratings, its pure tones and clicks, its taps and vibrations. These are techniques of analytic experiments and must not be regarded as revealing the initial stages, sensa, of the perceptual process, as if for example we sense a picture as a consequence of putting together small colour patches, as in a mosaic. On the contrary perceptions arise in consciousness as fully formed pictures or as sounds of speech or music, being replete with meaning (Gibson, 1966; Thorpe, 1974). It is not of course denied that there are subconscious processes on the way to the conscious perception of pictures or sounds or felt objects of infinite variety, as we shall discuss in Lecture 3.

The brain–mind problem has now come into an exciting era where conference after conference brings together philosophers, psychologists, psychiatrists, neurologists, neurobiologists and experts in machine intelligence. They declaim, they argue, they dispute and may even on rare occasions agree! Moreover there has been a rich variety of books as well as the numerous published texts of conferences. The book *The Self and Its Brain* by Popper and me (1977) would seem to have achieved the doubtful distinction of being at the centre of a philosophical storm, as witness the spate of critical reviews aimed at it. It is all to the good that there is this revival of interest in the greatest problem confronting us – our nature and our destiny. Perhaps it is good that such emotions are aroused by irreconcilable confrontations. But it is to be regretted that sometimes beliefs are held with dogmatic fervour, and that there may be a lack of humility in the uncritical zeal with which dogmatic beliefs are expounded. Before embarking on a consideration of brain–mind theories in all their diversity and even strangeness, I will consider a most important issue on which there are fascinating experimental investigations, namely our experience of mental unity or singleness.

1.2 Unity of Consciousness and Commissurotomy

It is a universal experience that subjectively there is a mental unity that is recognized as a continuity from one's earliest memories. It is the basis of the concept of the self. Weiss (1969) expresses this well, speaking of

> The unity which is my greatest experience: that even though I know I am constantly changing – all molecules are changing, everything in me is being turned over substantially – there is nevertheless my identity, my consciousness of being essentially the same that I was 20 years ago. However much I may have changed, the continuity of my identity has remained undisrupted.

Nevertheless we have two cerebral hemispheres and there has been an amazing range of discussion with respect to their functions and the degree to which they may exercise separate functions. Historically this disputation can be traced back to the identification of the speech centres in the left cerebral hemisphere (Fig. 1-1) by Broca and Wernicke as a result of locating the lesions that caused aphasias. There has subsequently been an enormous literature on the effects of clinical lesions on cerebral function, not only of the linguistic functions, but also for example of the apraxia from lesions of the right parietal lobe. Unfortunately clinical lesions are usually not well circumscribed and so do not give clear localization for the disordered function. A remarkable advance was made by Gordon Holmes who systematically investigated injuries by missiles in the First World War in order to define the topographic localization of the visual field in the occipital lobe (cf. Popper and Eccles, 1977, Fig. E2-5).

Most important information on the functions of the various lobes (Fig. 1-1) of both hemispheres has been derived from studies by Milner and others of patients in whom circumscribed lesions were

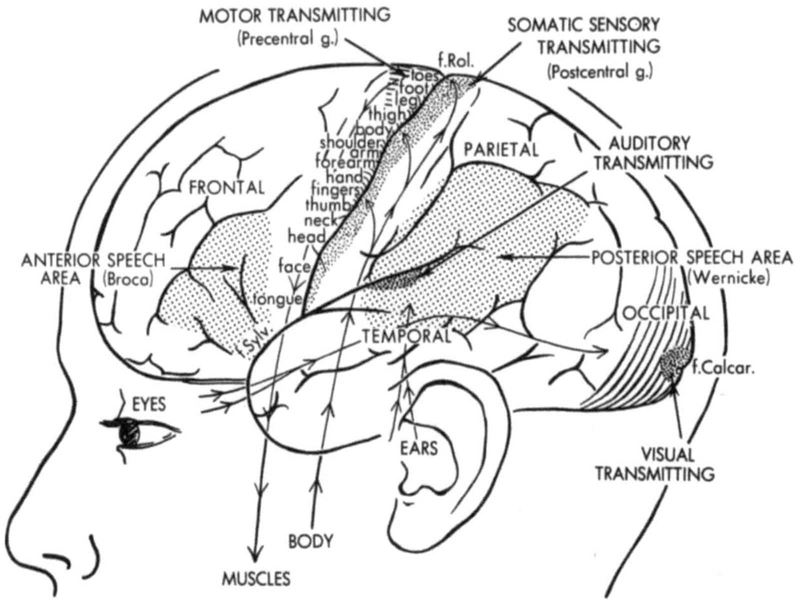

Fig. 1-1. The motor and sensory transmitting areas of the cerebral cortex. The approximate map of the motor transmitting areas is shown in the precentral gyrus, while the somatic sensory receiving areas are in a similar map in the postcentral gyrus. Actually the toes, foot and leg should be represented over the top on the medial surface. Other primary sensory areas shown are the visual and auditory, but they are largely in areas screened from this lateral view. The frontal, parietal, occipital and temporal lobes are indicated. Also shown are the speech areas of Broca and Wernicke

made for therapeutic reasons (cf. Chap. E6, Popper and Eccles, 1977). However by far the most important evidence relating to the unity of consciousness comes from the studies by Sperry and his associates on commissurotomized patients. In the operation there was section of the corpus callosum, the great tract of nerve fibres, about 200 million, that links the two cerebral hemispheres (Figs. 1-2, 1-3, 1-4). It must be recognized that the connections of the hemispheres to lower brain regions remain intact. It must further be recognized that the two hemispheres have been intimately linked in all the cerebral activities of the subject prior to the operation, and that each hemisphere will carry the memories of these many years of conjoint performances.

It is fortunate that severing this immense linkage by the 200 million fibres of the corpus callosum gives so little apparent disability

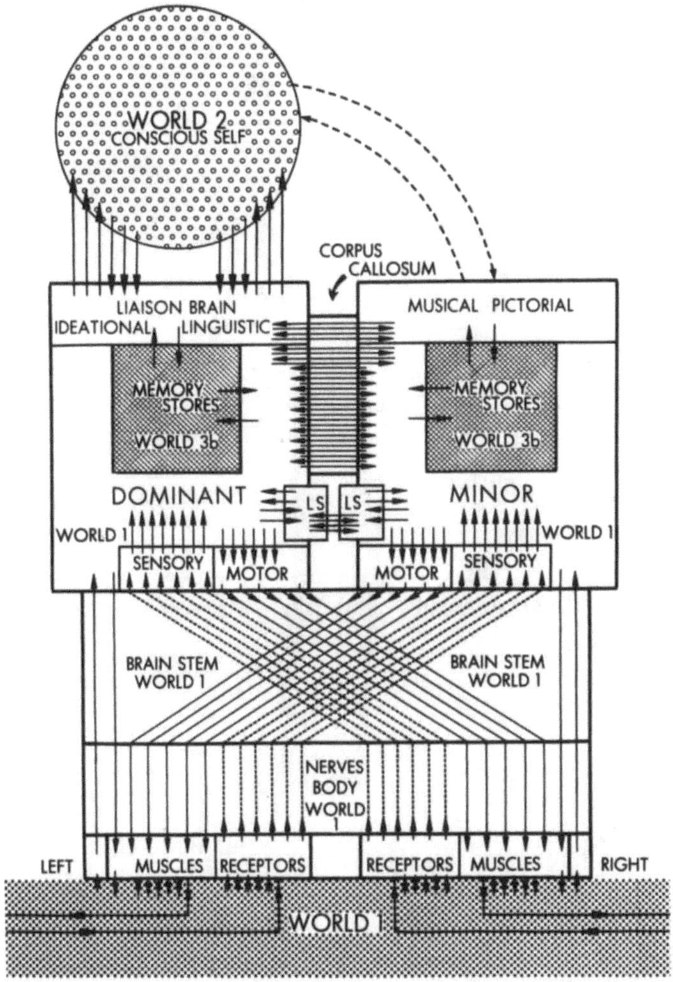

Fig. 1-2. Communications to and from the brain and within the brain. Diagram to show the principal lines of communication from peripheral receptors to the sensory cortices and so to the association areas of the cerebral hemispheres. Similarly, the diagram shows the output from the cerebral hemispheres via the motor cortex and so to muscles. Both these systems of pathways are largely crossed as illustrated, but minor uncrossed pathways are also shown by the vertical lines in the brain stem. The dominant left hemisphere and minor right hemisphere are labelled, together with some of the properties of these hemispheres that are found displayed in Fig. 1-5. The corpus callosum is shown as a powerful cross-linking of the two hemispheres, and, in addition, the diagram displays the modes of interaction between Worlds 1 and 2, as described in the text, and also illustrated in Figs. 1-7 and 7-1. Note the blocks marked *LS*, which represent the limbic system (cf. Lecture 5) with the ipsilateral and commissural connectivities

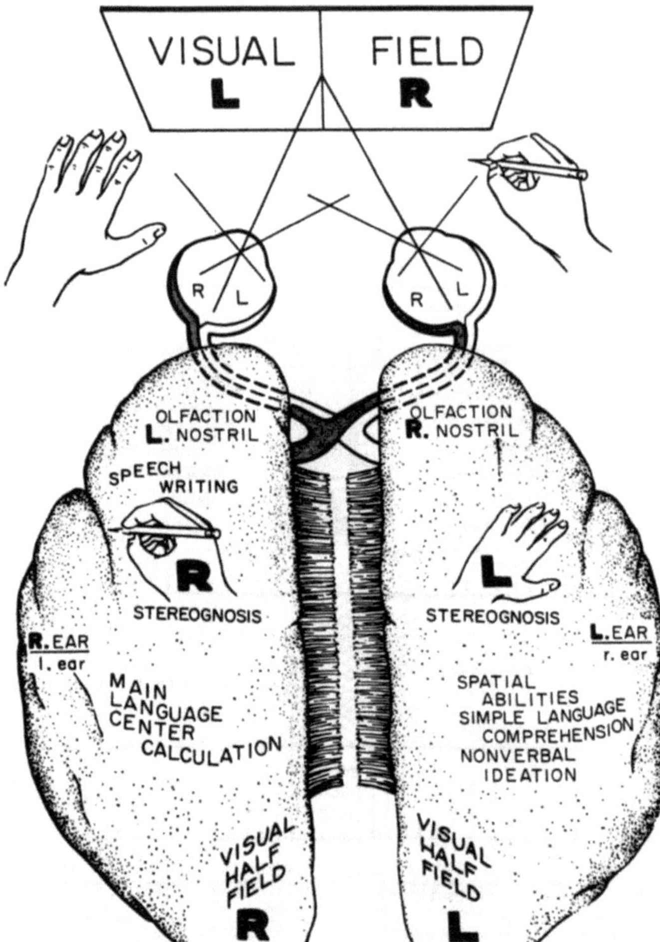

Fig. 1-3. Schema showing the way in which the left and right visual fields are projected onto the right and left visual cortices, respectively, due to the partial decussation in the optic chiasma (cf. Fig. 3-6). The schema also shows other sensory inputs from right limbs to the left hemisphere and that from left limbs to the right hemisphere. Similarly, hearing is largely crossed in its input, but olfaction is ipsilateral. The programming of the right hand in writing is shown pictorially to come from the left hemisphere. Sperry, R.W.: Lateral specialization in the surgically separated hemispheres. In: The neurosciences: third study program. Schmitt, F.O., Worden, F.G. (eds.), pp. 5–19. Cambridge (Mass): MIT Press 1974

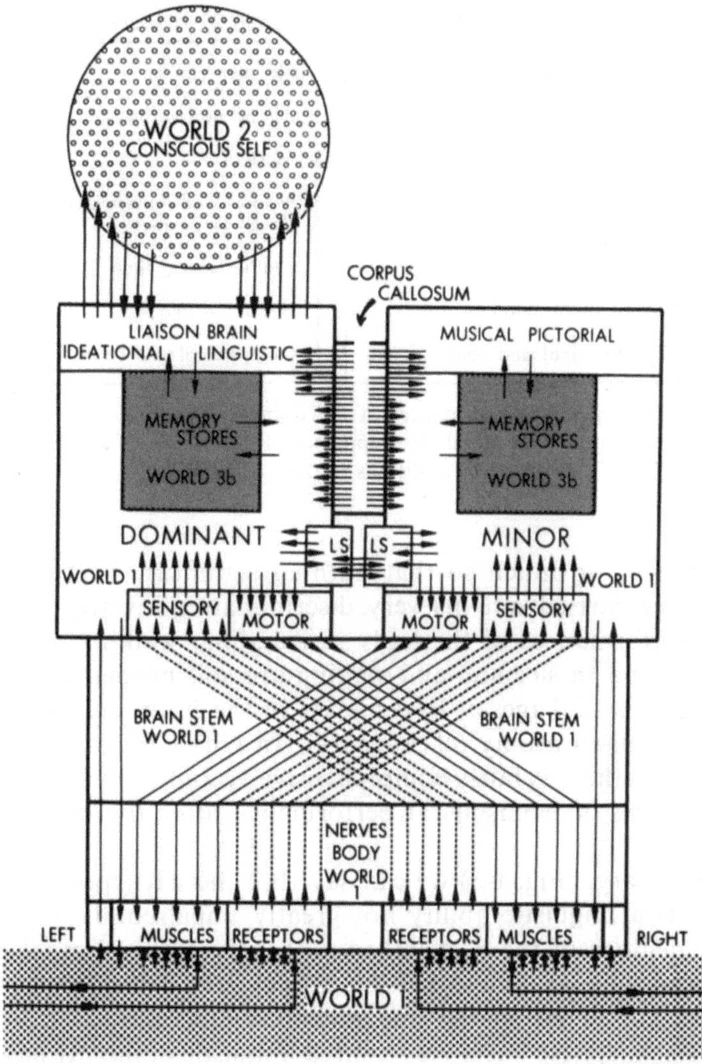

Fig. 1-4. Communications to and from the brain and within the brain as in Fig. 1-2, but after section of the corpus callosum. Communication from the brain to World 2 is now shown to be only from the dominant hemisphere. See text for further description

DOMINANT HEMISPHERE	MINOR HEMISPHERE
Liaison to self consciousness	No such Liaison
Verbal	Almost non-verbal
Linguistic description	Musical
Ideational Conceptual similarities	Pictoral and Pattern sense Visual similarities
Analysis over time	Synthesis over time
Analysis of detail	Holistic — Images
Arithmetical and computer-like	Geometrical and Spatial

Fig. 1-5. Various specific performances of the dominant and minor hemispheres as suggested by the new conceptual developments of Levy-Agresti and Sperry (1968) and Levy (1973). There are some additions to their original list

to the patient, so much so that nothing significant was recognized until Sperry carried out his very discriminative investigations. The general performances of the body in standing, walking, diving and swimming and in sleeping and waking are still normally linked because the cross-linkages at lower levels of the brain are not affected by the commissurotomy (cf. Fig. 1-4). In Chap. E5 of Popper and Eccles (1977) there is fairly comprehensive account of all the discriminative testing of hemispheral performance after the commissurotomy.

In summary, (Fig. 1-5) we can say that the left (speaking) hemisphere has a linguistic ability not greatly impaired. It also carries a good memory of the past linked with a good intellectual performance and with an emotional life not greatly disturbed. However it is deficient in all spatial and constructional tasks. By contrast the right hemisphere has a very limited linguistic ability. It has access to a considerable auditory vocabulary, being able to recognize commands and to relate words presented by hearing or vision to pictorial representations. It was also surprising that the right hemisphere responded to verbs as effectively as to action names. Despite all this display of language comprehension, the right hemisphere is extremely deficient in expression in speech or in writing, which is effectively zero. However, in contrast to the left hemisphere, it is very effective in all spatial and constructive tasks and it is also proficient in global recognition tasks.

Gordon et al. (1971) found that the unity of conscious experience was conserved after section of the anterior and middle thirds of the corpus callosum. It was concluded that the linkages of the parietal and occipital cortices in the posterior third are vitally concerned in conscious unity. However the commissural linkage between the frontal lobes must not be excluded because this linkage could be conserved by a transcallosal pathway that operates via the immense association connections to the parietal and temporal lobes and their intact callosal connections.

When more sophisticated investigations were developed, allowing up to 2 h of continued testing, it became clear that the right hemisphere was displaying evidence of conscious responses at a level superior to those exhibited by any non-human primates. It will be appreciated that there is no indubitable test for consciousness, but it is generally accepted that the higher animals, birds and mammals, display conscious behaviour when they act intelligently and emotionally and are able to learn appropriate reactions. On those criteria the consciousness of the right hemisphere is indubitable, as is diagrammed in Fig. 1-2. The perplexing question is whether the right hemisphere is self-conscious, meaning by that, that it knows and is conscious of its selfhood. As stated by Sperry et al. (1979):

> Self-consciousness appears to be almost strictly a human attribute, according to present evidence drawn mainly from mirror self-recognition tests. It seems not to be found in animals below the primates, and only to a limited extent in the great apes. In human childhood self-consciousness makes its appearance relatively late in development, appearing first at around 18 months of age. Thus ontogenetically as well as phylogenetically self-consciousness can be rated as a relatively advanced stage of conscious awareness.

Sperry et al. (1979) report investigations on two commissurotomy patients which were designed to test for aspects of self-consciousness and general social awareness in the right hemisphere. In these tests a wide variety of pictures of persons as well as of familiar objects and scenes were presented in assembled arrays to the left visual field of the patient and hence *exclusively* to the right hemisphere (cf. Fig. 1-3). The subject could always identify the familiar photograph in the ensemble of pictures, but there were difficulties in specifying what it was and the investigators had to adopt a rather informative prompting system before the right hemisphere identification could be expressed in language, presumably by the left hemisphere.

Their dramatic conclusions of approximate equality of the two hemispheres in identification may be criticized as being derived from a rather optimistic over-interpretation of the subject's responses, as illustrated in the experimental protocols. Nevertheless there is remarkable evidence in favour of a limited self-consciousness of the right hemisphere. I quote the most convincing account of the last part of a testing session in which the subject LB had been expressing his approval or disapproval of various photographs of persons by thumbs up or thumbs down signalling by his left hand.

> Towards the end of this testing session, LB was presented with a choice array containing 4 portrait photos of adult males, 3 strangers and one of himself in the lower left position. When asked if he recognized any of these LB pointed promptly to the photo of himself. Asked for a thumb sign evaluation, he gave a decisive 'thumbs-down' response but unlike other 'thumbs-down' signals, this one was accompanied by a wide, sheepish and (to all appearances) a self-conscious grin. When we then asked if he knew who it was, LB after only a short hesitation guessed correctly 'myself'. *Interpretation:* LB recognized himself readily with the right hemisphere. The tongue-in-cheek 'thumbs-down' response to his own photo accompanied by a broad grin indicates not only self-recognition in the minor hemisphere but also a subtle sense of humor and self-conscious perspective befitting the total situation. The emotional effect was transferred centrally and also peripherally and was sufficiently distinctive, combined with other cues, that the left hemisphere soon guessed the correct identification.

It can hardly be doubted that the right hemisphere has at least a limited self-consciousness.

In summary Sperry et al. (1979) state:

> The ability of the commissurotomy subjects with visual input lateralized to the left half field to recognize, select and identify from among neutral items in a choice array pictures of themselves, their family, relatives, acquaintances, pets, belongings and also political, historical and religious figures and personalities from the entertainment world, all at a level quite comparable to that of the left hemisphere of the same subject is taken to indicate the presence in the right hemisphere of a well developed sense of self and social awareness. The kinds of emotional reactions that were generated and the selectivity of responses to follow-up questions of the examiners and to vocal cues from the subjects' own comments showed that true identifications were made in the right hemisphere. ... It was possible to exclude significant assistance from the vocal hemisphere in the initial identification process in most instances because the content of the vocal comments indicated that the speaking hemisphere had remained unaware of what the mute hemisphere had recognized and was reacting to.
>
> The overall level of the right hemisphere's ability to identify test items and also the quality of the accompanying emotional and evaluative responses were of

> the same order approximately as those obtained from the right visual field and left hemisphere. Occasional discrepancies between left field and right field responses were the exception rather than the rule, and did not exceed the intrahemispheric range of variation from one test session to another and in general can hardly be considered indicative of valid left-right differences. Taken together, the present data strongly reinforce the assumption that human subjectivity is basically much the same in the two hemispheres.

Actually the emotional reactions generated by picture stimulation of the right hemisphere would be dependent on the connectivities to the limbic system in particular (cf. Lecture 5), which are unaffected by commissurotomy and so are able to spread to the left hemisphere to give a consciously experienced emotion (cf. Fig. 1-4).

These tests for the existence of mind and of self-conscious mind are at a relatively simple pictorial and emotional level. We can still doubt if the right hemisphere has a full self-conscious existence. For example, does it plan and worry about the future, does it make decisions and judgements based on some value system? These are essential qualifications for personhood as ordinarily understood (Strawson, 1959; Popper and Eccles, 1977, Sects. 31 and 33). Let us now consider the bearing of these findings on the unity of the consciously experiencing self.

As pointed out by DeWitt (1975), very important problems are raised by the commissurotomy investigations of Sperry and associates and a wide range of interpretations has been offered.

(1) Puccetti (1973) is the most extreme in that he regards the commissurotomy patients as having two minds with radically different properties, and each of these is a distinct person. In fact he goes so far as to propose that this dual personality obtains before the commissurotomy. I quote his actual statements:

> Either commissurotomy results in two minds or it does not; if it does, it also yields two persons.
> But even this last statement is misleading. How can commissurotomy *create* two minds or persons if there was just one before? Which mind, the left brain-based one or the right brain-based one is brand new? And how are we to make a choice here? Both brains, as we have seen, were conscious and functioning in their rather specialised ways before the operation. It is just that they then functioned more synchronously – because of the commissural connections... If that is so, if we cerebrally intact twin-brained human beings are really compounds of two persons, which is me? Am I the person whose conscious unity is rooted in left brain information-processing and right hand motor control; or am I the person whose consciousness is based in right brain activity and subordinate left hand control?

(2) Sperry (1976), Bogen (1969) and Gazzaniga (1971) also propose that there are two self-conscious minds, i.e., the commissurotomy has split the mind into two, even the self-conscious mind, but normally there is only one person. (Bogen, 1969) states:

> One of the most obvious and fundamental features of the cerebrum is that it is double. Various kinds of evidence, especially from hemispherectomy, have made it clear that one hemisphere is sufficient to sustain a personality or mind. We may then conclude that the individual with two intact hemispheres has the capacity for two distinct minds, – each of us has two minds in one person.

(3) The view of MacKay (1978) and Eccles (1979a) is that in commissurotomy a conscious mind has been split off in association with the right hemisphere, the left hemisphere remaining with a relatively intact self-conscious mind and the associated personhood.

DeWitt (1975) makes the point that the presence of language marks the difference between the existence of *self-consciousness* and the complete absence of any awareness of self. I would agree in general, but I worry about what is meant by language because the right hemisphere certainly has a fair performance in understanding language though it is at zero level in expression (Zaidel, 1976). I think that, in the light of Sperry et al.'s (1979) recent investigations, there is some self-consciousness in the right hemisphere, but it is of a limited kind and would not qualify the right hemisphere to have personhood on the criteria mentioned above. Thus the commissurotomy has split a fragment off from the self-conscious mind, but the person remains apparently unscathed with mental unity intact in its now exclusive left hemisphere association. However, it would be agreed that emotional reactions stemming from the right hemisphere can involve the left hemisphere via the partly unsplit limbic system (Fig. 1-4). So the person remains emotionally attached also to the right hemisphere.

I would agree with DeWitt's (1975) interpretation of the situation after commissurotomy:

> Both minor and major hemispheres are conscious in that they both, no doubt, have the basic phenomenal awareness of perceptions, sensations, etc. And they both have minds ... in that they exhibit elaborated, organised systems of response hierarchies, i.e., intentional behaviour. But in addition I would conjecture that only the major hemisphere has a self; only the language utilising brain is capable of the abstract cognising necessary in order to be aware of itself as a unique being. In a word, only the major hemisphere is aware of itself *as a self*.

This corresponds to the situation in real life, where the associates of the patient find no difficulty after the operation in regarding it as the self or person that it was before the operation. The patients themselves would of course concur, but they do have a problem arising from the splitting of the conscious mind. There is the difficulty in controlling the movements emanating from the activity of the right hemisphere with its associated mind. These movements are completely beyond the control of the conscious self or person that is exercised through the left hemisphere. For example they refer to their uncontrollable left hand as their 'rogue hand'.

It would seem that this interpretation of DeWitt conforms with all the observational data on the commissurotomy subjects, but avoids the extreme philosophical difficulties inherent in the hypothesis of Puccetti that even normally there is a duality of personhood – 'two persons in one brain' as he provocatively expresses it.

Nagel (1971) has presented a critical study of the findings on commissurotomized patients in an attempt to arrive at a solution of the numbers of minds and numbers of persons! He finds it difficult to conceive what it may be like to be a commissurotomy patient. He favours the hypothesis that they have two minds, but finds this in conflict with the behavioural integration displayed by the patients. So he finishes with a destructive critique of the whole concept of the person and retreats to some behavioural control system within the brain and between the hemispheres that somehow unites all the complexities of the individual person. But he regards the unity of the person as an illusion!

One recognizes the difficulties raised by the splitting of the mind in commissurotomy and even more so in the existence of a fragment of the self-conscious mind associated with the isolated right hemisphere, but this is no excuse for Nagel's pessimism, as he calls it! We can agree with his statement in footnote 11 that commissurotomy is just one way of dividing the brain. There could be cortical deconnections within one hemisphere, separating the prefrontal lobes, as was bilaterally done in prefrontal leucotomy for example. But, as we shall see in Lecture 2 the liaison of mind with brain must not be treated as a problem of limited regions of action. It will be proposed that the self-conscious mind normally is in liaison with an immense territory of cortical modules that are dispersed through the whole neocortex. Furthermore it has to be recognized that there is an immense ongoing operation of communication between modules by impulses in the association fibres of one hemisphere numbering about two thousand million as well as the 200 million commissural fibres. It is this immense communication system that enables the spatio-temporal patterns of activity in the 4 million modules (cf. Lecture 2) to give in coded form the extremely rich information that is read out by the self-conscious mind and that is the basis of our ineffable experience of mental unity.

1.3 Hypotheses Relating to the Brain–Mind Problem

1.3.1 Introductory Considerations

It is not possible here to give a detailed appraisal of the immense philosophical literature on the brain–mind or body–mind problem. Fortunately this has recently been done in a masterly manner by Popper (Popper and Eccles, 1977, Chaps. P1, P3, P4, and P5). He has critically surveyed the historical development of the problem from the earliest records of Greek thought. I will begin by a simple description and diagram of the principal varieties of this extremely complex and subtle philosophy, concentrating specifically on the formulations that relate to the brain rather than the body, because clinical neurology and the neurosciences make it abundantly clear that the mind has no direct access to the body. All interactions with the body are mediated by the brain, and furthermore only by the higher levels of cerebral activity.

For our present purpose it is of value to clarify the arguments firstly by introducing the three-world philosophy of Popper (cf. Fig. 7-1) and secondly by developing an explanatory diagram (Fig. 1-6) of the principal theories so that the materialist theories of the mind can be contrasted with the dualist-interactionist theory that is here being proposed.

World 1 is the whole material world of the cosmos both inorganic and organic, including all of biology, even human brains, and all man-made objects.

World 2 is the world of conscious experiences, or of the mind, not only of our immediate perceptual experiences, visual, auditory, tactile, pain, hunger, anger, joy, fear etc., but also of our memories, imaginings, thoughts, planned actions, and centrally thereto of our unique self as an experiencing being.

World 3 is the world of human creativity – for example the objective contents of thoughts underlying scientific, artistic and literary expression. Thus World 3 is the world of culture in all of its manifestations, as has been expressed by Popper in Chap. P2 of Popper and Eccles (1977).

1.3.2 Survey of Brain–Mind Hypotheses

The dominant theories of the brain–mind relationship that are today held by neuroscientists are purely materialistic in the sense that the brain is given complete mastery (Pribram, 1971; Rensch, 1971, 1974; Barlow, 1972; Doty, 1975; Wilson, 1976; Blakemore, 1977; Mountcastle, 1978b; Edelman, 1978). The existence of mind or consciousness is not denied, but it is relegated to the passive role of mental experiences accompanying some types of brain action, as in psychoneural identity, but with absolutely no *effective* action on the brain. The complex neural machinery of the brain functions in its determined materialistic fashion regardless of any consciousness that may accompany it. The 'common sense' experiences that we can control our actions to some extent or that we can express our thoughts in language are alleged to be illusory. Actually it is rare for this to be stated so baldly, but despite all the sophisticated cover-up the situation is exactly as stated. An effective causality is denied to the self-conscious mind *per se*.

In Fig. 1-6 World 1 is divided into World 1_P and an infinitesimally small World 1_M. In general, materialist theories are those subscribing to the statement that mental events can have no *effective* action on the brain events in World 1 – that World 1 is closed to any conceivable outside influence such as is postulated in dualist-interactionism. This closedness of World 1 is ensured in four different ways in the four varieties of materialism illustrated in Fig. 1-6.

1) *Radical materialism:* It is asserted that all is World 1_P. There is a denial or repudiation of the existence of mental events. They are simply illusory. The brain–mind problem is a non-problem!

2) *Panpsychism:* It is asserted that all matter has an inside mental or protopsychical state. Since this state is an integral part of matter, it can have no action on it. The closedness of World 1 is safeguarded.

3) *Epiphenomenalism:* Mental states exist in relation to some material happenings, but causally are completely irrelevant. Again the closedness of World 1 is safeguarded.

Diagrammatic representation of brain–mind theories

World 1 = All of material or physical world including brains
World 2 = All subjective or mental experiences
World 1_P is all the material world that is without mental states
World 1_M is that minute fraction of the material world with associated mental states

Radical Materialism:	World 1 = World 1_P; World 1_M = 0; World 2 = 0.
Panpsychism:	All is World 1–2, World 1 or 2 do not exist alone.
Epiphenomenalism:	World 1 = World 1_P + World 1_M World $1_M \rightarrow$ World 2
Identity theory:	World 1 = World 1_P + World 1_M World 1_M = World 2 (the identity)
Dualist – Interactionism:	World 1 = World 1_P + World 1_M World $1_M \rightleftarrows$ World 2; this interaction occurs in the liaison brain, LB = World 1_M. Thus World 1 = World 1_P + World 1_{LB}, and World $1_{LB} \rightleftarrows$ World 2

Fig. 1-6. Schematic representation of the various theories of brain and mind. Full description in text

4) *The identity theory or the central state theory or the psychoneural identity theory:* Mental states exist as an inner aspect of some material structures that in present formulations are restricted to brain structures such as nerve cells. This postulated 'identity' may appear to give an effective action, just as the 'identical' nerve cells have an effective action. However the result of the transaction is that the purely material events of neural action are themselves *sufficient* for all brain–mind responses, hence the closedness of World 1 is preserved. This has been very well argued by Beloff (1976).

An outstanding characteristic of identity theories that centre around Feigl's (1967) brilliant formulation is the multiplication of names for theories that are almost indistinguishable. Not only are there the three names given above, but there are for example: emergent interactionism (Sperry, 1976, 1977); identistic panpsychism (Rensch, 1971); physicalism (Smart, 1963, 1978); biperspectivism (Laszlo, 1972); emergentistic materialism (Bunge, 1977).

In contrast to these materialist or parallelist theories are the *dualist-interaction* theories, as diagrammed at the bottom of Fig. 1-6. The essential feature of these theories is that mind and brain are independent entities, the brain being in World 1 and the mind in World 2, and that they somehow interact, as illustrated by the arrows

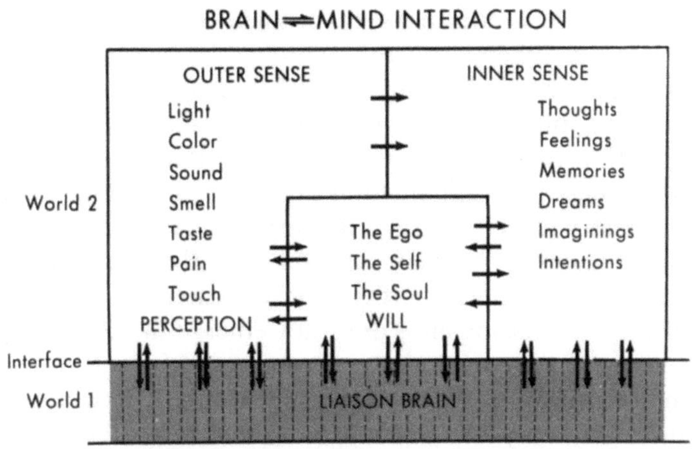

Fig. 1-7. Information flow diagram for brain–mind interaction. The three components of World 2: outer sense, inner sense and the ego or self are diagrammed with their communications shown by *arrows*. Also shown are the lines of communication across the interface between World 1 and World 2, that is from the liaison brain to and from these World 2 components. The liaison brain has the columnar arrangement indicated by the *vertical broken lines*. It must be imagined that the area of the liaison brain is enormous, with open modules numbering over a million, not just the two score here depicted. Full description in text

in Fig. 1-7. Thus there is a frontier, as diagrammed in Fig. 1-7, and across this frontier there is interaction in both directions, which can be conceived as a flow of information, not of energy. Thus we have the extraordinary doctrine that the world of matter-energy (World 1) is not completely sealed, which is a fundamental tenet of physics, but that there are small 'apertures' in what is otherwise the completely closed World 1. On the contrary, as we have seen, closedness of World 1 has been safeguarded with great ingenuity in all materialist theories of the mind. Yet I shall now argue that this is not their strength, but instead their fatal weakness (cf. Popper and Eccles, 1977).

1.3.3 Critical Evaluation of Brain–Mind Hypotheses

Great display is made by all varieties of materialists that their brain–mind theory is in accord with natural law as it now is. However, this claim is invalidated by two most weighty considerations.

Firstly, nowhere in the laws of physics or in the laws of the derivative sciences, chemistry and biology, is there any reference to consciousness or mind. Shapere (1974) makes this point in his strong criticisms of the panpsychist hypothesis of Rensch (1974) and Birch (1974) in which it was proposed that consciousness or protoconsciousness is a fundamental property of matter. Regardless of the complexity of electrical, chemical or biological machinery there is no statement in the 'natural laws' that there is an emergence of this strange non-material entity, consciousness or mind. This is not to affirm that consciousness does not emerge in the evolutionary process, but merely to state that its emergence is not reconcilable with the natural laws as at present understood. For example such laws do not allow any statement that consciousness emerges at a specified level of complexity of systems, which is gratuitously assumed by all materialists except panpsychists. Their belief that some primordial consciousness attaches to all matter, presumably even to atoms and subatomic particles (Rensch, 1971), finds no support whatsoever in physics. One can also recall the poignant questions by computer-lovers. At what stage of complexity and performance can we agree to endow computers with consciousness? Mercifully this emotionally charged question need not be answered. You can do what you like to computers without qualms of being cruel!

Secondly, all materialist theories of the mind are in conflict with biological evolution. Since they all (panpsychism, epiphenomenalism, and the identity theory) assert the causal ineffectiveness of consciousness *per se*, they fail completely to account for the biological evolution of consciousness, which is an undeniable fact. There is firstly its emergence and then its progressive development with the growing complexity of the brain. In accord with evolutionary theory only those structures and processes that significantly aid in survival are developed in natural selection. If consciousness is causally impotent, its development cannot be accounted for by evolutionary theory. According to biological evolution mental states and consciousness could have evolved and developed *only if they were causally effective* in bringing about changes in neural happenings in the brain with the consequent changes in behaviour. That can occur *only if* the neural machinery of the brain is open to influences from the mental events of the world of conscious experiences, which is the basic postulate of dualist-interactionist theory. As Sherrington (1940) states:

> The influence of mind on the doings of life makes mind an effective contribution to life. We can seize then how it is that mind counts and has counted. That it has been evolved seems to assure us that it has counted. How it has counted would seem to be that the finite mind has influenced its individual's 'doing'.

Finally the most telling criticism of all materialist theories of the mind is against its key postulate that the happenings in the neural machinery of the brain provide *a necessary and sufficient explanation of the totality both of the performance and of the conscious experience of a human being.* For example the willing of a voluntary movement is regarded as being *completely determined* by events in the neural machinery of the brain, as also are all other cognitive experiences. But as Popper states (Popper, 1972, Chapt. 6).

> According to determinism, any theory such as say determinism is held because of a certain physical structure of the holder – perhaps of his brain. Accordingly, we are deceiving ourselves and are physically so determined as to deceive ourselves whenever we believe that there are such things as arguments or reasons which make us accept determinism. In other words, physical determinism is a theory which, if it is true, is unarguable since it must explain all our reactions, including what appear to us as beliefs based on arguments, as due to purely physical conditions. Purely physical conditions, including our physical environment make us say or accept whatever we say or accept.

This is an effective *reductio ad absurdum.* This stricture applies to all of the materialist theories. So perforce we turn to dualist-interactionist explanations of the brain–mind problem, despite the extraordinary requirement that there be effective communication in both directions across the frontier shown in Fig. 1-7.

Necessarily the dualist-interactionist theory is in conflict with present natural laws, and so is in the same 'unlawful' position as the materialist theories of the mind. The differences are that this conflict has always been admitted and that the neural machinery of the brain is assumed to operate in strict accordance to natural laws except for its openness to World 2 influences.

Moreover, as stated by Popper (Popper and Eccles, 1977, Dialogue XII), the interaction across the frontier in Fig. 1-7 need not be in conflict with the first law of thermodynamics. The flow of information into the modules could be effected by a balanced increase and decrease of energy at different but adjacent micro-sites, so that there was no net energy change in the brain. The first law at this level 'may be valid only statistically'.

1.4 Recent Proposals by Neuroscientists in Relation to the Brain–Mind Problem

After this introductory survey I turn to some recent theoretical developments that have been proposed in relation to scientific studies on the brain. We can assimilate these various theories into the framework of Fig. 1-6.

Sherrington was the pioneer in the endeavour to reinstate the conscious mind as an effective agent in controlling brain activity, and so to contradict the basic beliefs of behaviourism and scientific materialism. Already in his Rede Lecture (Sherrington, 1933), he had proposed the dualism of mental experience and brain and their interaction. In his Gifford Lectures of 1937 and 1938, here in Edinburgh, he made this brain–mind problem the central theme, building up and elaborating a great structure of dualist-interactionism (Sherrington, 1940; Eccles and Gibson, 1979, Chap. 7).

Later I developed Sherrington's ideas especially in relation to the known anatomical and physiological properties of the neocortex (Eccles, 1951b, 1953).

As early as 1952, Sperry had published on the relationship of brain to mind, and from 1965 to the present time he has progressively developed a brain–mind theory that he calls *emergent-interactionism*. This theory has some very subtle aspects that I and others (e.g. Popper) find elusive. Nevertheless, it was of great importance that a brain scientist with a deservedly great reputation should develop an hypothesis that was in conflict with the materialist monism of the behaviourist establishment. I will discuss his theory in relation to its expression in one of his most recent publications (Sperry, 1976) and give quotations therefrom. In a more recent paper (Sperry, 1977) there is a similar account.

The essential postulates of his theory are initially the orthodox anatomy and physiology of the neocortex with their emergent phylogenetic development. On this basis he postulates an operational derivation of the conscious properties which are assumed to have causal potency in regulating the course of the brain events. He states that

'the subjective conscious experience becomes an integral part of the brain process' and that

'the mental events are *causes* rather than *correlates*. In this respect our view can be said to involve a form of mental interactionism, except that there is no implication of dualism or other parallelism in the traditional sense. The mental forces are direct causal emergents of the brain process'.

But then he goes on to speak of

'the holistic conscious properties ... of the brain process. These special mental properties have not been described objectively as yet in any form. They are holistic configurational properties that have yet to be discovered. We predict that, once they have been discovered and understood, they will be best conceived of as being different from and more than the neural events of which they are composed'.

It seems that Sperry is here postulating that there are holistic configurational properties of mental events that are composed of neural events of a lower order. The dilemma arises because he avoids dualism by not giving the mental events an independence from neural events, yet at the same time he gives the mental events a causal influence on the neural events in a typical dualist-interactionist manner, stating that 'the brain physiology determines the mental effects and mental phenomena in turn have causal influence on the neurophysiology'. I find it difficult to accept Sperry's rejection of dualism, because this rejection seems in conflict with his proposal of the most important operational considerations on brain–mind interaction and the derivations therefrom. There will be further references to his theory in Lectures 6 and 7. Already there has been consideration of his theory of emergent interactionism in relation to problems of consciousness in the hemispheres of the commissurotomized brain. The rejection of dualism leads to the only alternative position, namely a very sophisticated identity hypothesis (cf. Fig. 1-6) relating to the

> emergent holistic properties of high order cerebral processes, and further that these emergent phenomena will be seen to play a potent causal role in brain function that cannot be accounted for in terms merely of the neurophysiologic and neurochemical events as these are traditionally conceived (Sperry, 1976).

I wonder if the spatio-temporal patterning of modular actions that are now being proposed (Szentágothai, 1978a, b, 1979; Eccles, 1979a; Chaps. 2, 3 and 4 below) would qualify as the emergent holistic properties of the neocortex.

Very recently a theory of emergentist materialism has been proposed with great display (Bunge, 1977; Bunge and Llinás, 1978) that in most features is almost the same as the emergent interactionism just described that Sperry has been proposing since 1965. Unfortunately these authors make no reference to Sperry. It is also regrettable that their attack on dualist-interactionism reveals that they are ignorant of its essential features. Bunge and Llinás (1978) apparently believe that according to dualist-interactionism the mind is *independent* of the brain and not in interaction with it! Otherwise how could they ask?

> How is it possible that an aloof nonphysical mind can be so mercilessly demurred by such a relatively simple molecule as the lysergic acid diethyl amide? How is it that physical parameters which change the functional properties of the brain change the faculties of the mind?

They are refuting an extraordinary dualist-interactionism of their own invention! And again (Bunge, 1977):

> Unlike dualism, which digs an unbridgeable chasm between man and beast, emergentist materialism *jibes with evolutionary biology*, which – by exhibiting the gradual development of the mental faculties along certain lineages – refutes the superstition that only Man has been endowed with a mind.

Contrast this arrogant claim with the open discussion in the preceding section on the Unity of Consciousness and Commissurotomy; as well as many later discussions in these lectures. Here is an example of the extravagant overselling that Bunge (1977) indulges in:

> emergentist materialism is the only philosophy of mind that enables a breakthrough in the scientific investigation of the mind–body (or rather brain–rest of the body) problem – and that defends neuroscience against obstruction by obsolete philosophies and idealogies.

All this is very impressive for the layman, but why write this in a scientific publication that is distributed to neuroscientists who have been getting on quite well these many years before Bunge arrived in 1977? Nevertheless there are some redeeming features amongst all this pomposity:

> For example, rather than say that love can color our reasonings, we may say that the right brain hemisphere affects the left one, and that sex hormones can act upon the cell assemblies that do the thinking.

This seems to suggest that Bunge gets all tangled up with love, and why omit the hypothalamus? And so we leave him, but reference can be made to MacKay's (1978) criticism and devaluation.

On analogy with information systems engineering MacKay (1978) proposes that

> the direct correlate of conscious experience is the self-evaluating supervisory or 'metaorganizing' activity of the cerebral system that determines norms and priorities and organizes the internal state of readiness to reckon with the sources of sensory stimulation. On this view, although all the modules of the cerebral cortex could normally participate in the loops of information-flow to and from the supervisory system, their activity in isolation would not be the direct correlate of our conscious experience. This view would take the concept of conscious human agency as primary, and recognize an irreducible duality in the two kinds of correlated data we have about it: (a) the data of our conscious experience, and (b) the data obtained from observation of our brain workings. The essential condition is that our conscious form-determining activity must be embodied in, rather than running alongside, the brain activity that physically determines (at the energetic level) what our bodies do.

I have much sympathy with this effort of MacKay to preserve the causal effectiveness of human agency without coming into conflict with the first law of thermodynamics, but I fail to understand it in terms of the properties of neurones and neurone assemblages. What is the neural status of the postulated 'supervisory system', and where is it located?

It is surprising to find that panpsychism is still alive. Rensch (1971, 1974) has developed this philosophy at great length and with a wealth of detail from his knowledge of biology and Birch (1974) also expressed a panpsychist philosophy. Rensch repeatedly states that dualist-interactionism is contrary to the fundamental laws of physics, but, as pointed out above, Shapere (1974) criticizes Rensch's hypothesis on the grounds of its incompatibility with physics. Rensch (1971, p. 272) expresses an extraordinary panpsychism where he considers 'the evolution of consciousness from protopsychical elements to self-consciousness' by a series of stages from elementary particles up eventually to higher organisms with 'sensation, memory, stream of consciousness'.

The varieties of identity theory expressed by Barlow (1972) and Doty (1975) have already been described (Popper and Eccles, 1977, Chap. E7) and conform to the brief description given above in relation to Fig. 1-6. Their views may be taken as typical of the great majority of neuroscientists (cf. Mountcastle, 1978b). Uttal (1978) has given an authoritative account of the brain–mind problem that is based on the identity theory of Feigl. He flatly rejects dualist-interactionism.

Creutzfeldt and Rager (1978) base a dualist interpretation of the problems relating consciousness to brain events on a sophisticated philosophical analysis that stems from Kant and Husserl. They do not consider in detail the neural events that could be related to mental events, but make some important general statements:

> Our reason cannot accept any attempt to reduce experience of the human self to neurophysiological facts and to establish a monism of information.
> If neurophysiology as a science cannot explain phenomena such as perception, experience, consciousness or free will, we ask how indeed they can be envisaged. The only way of looking at them is to reflect upon the context in which they appear.
> We should always keep in mind that even a complete knowledge of the connectivities and operation of the brain – if this is at all possible – cannot explain the phenomena of our life experience. Even the most complete scientific image of our conscious experience is bound to be a mere image and not the experience itself.

These general statements will be of great value as we explore various aspects of the human psyche in these lectures.

Lorenz (1977) has also presented valuable insights into the brain–mind problem – or body–soul problem as he calls it:

> What matters here is solely the fact that the gulf between the physical and the spiritual is of a fundamentally different kind from that between the organic and the inorganic, and that between man and the animals.These latter gulfs represent transitions, and each owes its existence to a unique event in the evolution of the world. Not only can both be bridged by a theoretical continuum of intermediate forms, but we also know that these transitional forms have indeed existed at specific moments in time.... The 'hiatus' between soul and body, on the other hand, is indeed unbridgeable, albeit perhaps 'only for us', – ... that is, with the cognitive apparatus at our disposal. Yet I do not believe that this is a limitation imposed just by the present state of our knowledge, or that even a utopian advance of this knowledge would bring us closer to a solution of the problem. The autonomy of personal experience and its laws cannot in principle be explained in terms of chemical and physical laws or of neuro-physiological structure, however complex.

Here is a clear recognition of the duality of mind and matter. There are some neuroscientists (Kety, 1969; Thorpe, 1974; Penfield, 1975) who subscribe to a dualist-interactionist philosophy such as that expressed by Popper and Eccles (1977). Thorpe in particular has presented a philosophy almost identical with that which I will be developing in Lecture 2.

Lecture 2

Modules of the Neocortex and Their Role in Dualist-Interactionism

Résumé

An introduction to the essential neuronal structure of the cerebral cortex, leads on to an account of its *modular structure* as revealed by radiotracer and degeneration techniques. Each module is a column of a few thousand nerve cells that is vertically oriented across the cerebral cortex, being about 0.25 mm across and 2 to 3 mm long. These modules are now recognized to be the functional units of communication throughout the *association cortex*, which forms 95% of the human neocortex. It is therefore of the greatest importance to consider the manner in which these modular units, about four million in number, contribute to the total performance of the human cerebrum.

It is argued that, with this number of modules and with four parameters of modular operation as variants, the cerebral cortex has the capacity to generate virtually infinite numbers of spatio-temporal patterns with modules as units – sufficient to be the unique material substrates of a whole life-time of experiences and memories. An analogy with piano music is developed, where with 88 keys and using a similar four parameters as variants the whole of piano music has been constructed. A human brain can be thought of as a piano with 4 million keys.

These insights into the *pattern-generating capability of the cortical modules* lead to new concepts of the interaction between the modular patterns in the cortex and the self-conscious mind of World 2. It is conjectured that, dependent on the activity of the modules, they may be open or closed to interaction with the self-conscious mind. An illustrative diagram is presented showing spatial patterns of modules, open, closed and half open. It will be the basic reference diagram for the subsequent lectures.

2.1 Introduction

Since the greater part of this lecture series on the human psyche will be based on our scientific understanding of the human brain and particularly of the neocortex, it is desirable to give a very simple account of the way in which the neocortex is built of its constituent neurones and the manner in which these neurones communicate in order to provide the basis for the integrated performance of the brain at its highest level.

As diagrammed in Fig. 1-7 the essential feature of the hypothesis of dualist-interactionism is the active role of the self-conscious mind in its relationship to the neuronal machinery of special regions of the brain. The cerebral hemispheres are the most recently evolved parts of the brain and are distinguished by the great folded cortex of the cerebral hemispheres that is about 2.5 mm thick and 2500 cm^2 in area (cf. the left hemisphere in Fig. 1-1). A wide variety of experimental investigations and clinical observations establishes that this cerebral neocortex is the part of the brain especially concerned with conscious experiences; hence we have to concentrate on its structure and its function when we are attempting to discover the unique properties of the brain which give it this transcendent role of being in liaison with the mental events of conscious experiences. Special regions of the cerebral neocortex are represented as the liaison brain in Fig. 1-7, which presents an interface to the array of World 2 components.

The cerebral cortex is composed of about ten thousand million nerve cells or neurones, which are the basic biological units. Each has a central zone or body (soma) about 1/100 to 1/50 mm across with tree-like branches, dendrites, and a longer thin filament or axon that conveys messages or impulses from the neurone to other neurones; the largest and most numerous neurones of the neocortex have a pyramidal shape, hence being called pyramidal cells. These neurones are so densely packed that the individual neurone can be recognized in histological sections only when it is picked out by the extraordinarily fortunate staining procedure discovered by Golgi. For example for Fig. 2-1A only about 1% of the neurones were stained. Several pyramidal cells can be recognized with the branching tree-like dendrites and the thin axon (nerve fibre) projecting downwards from the centre of the soma or body. In Fig. 2-1B there is shown one such pyramidal cell with short spines (s) on

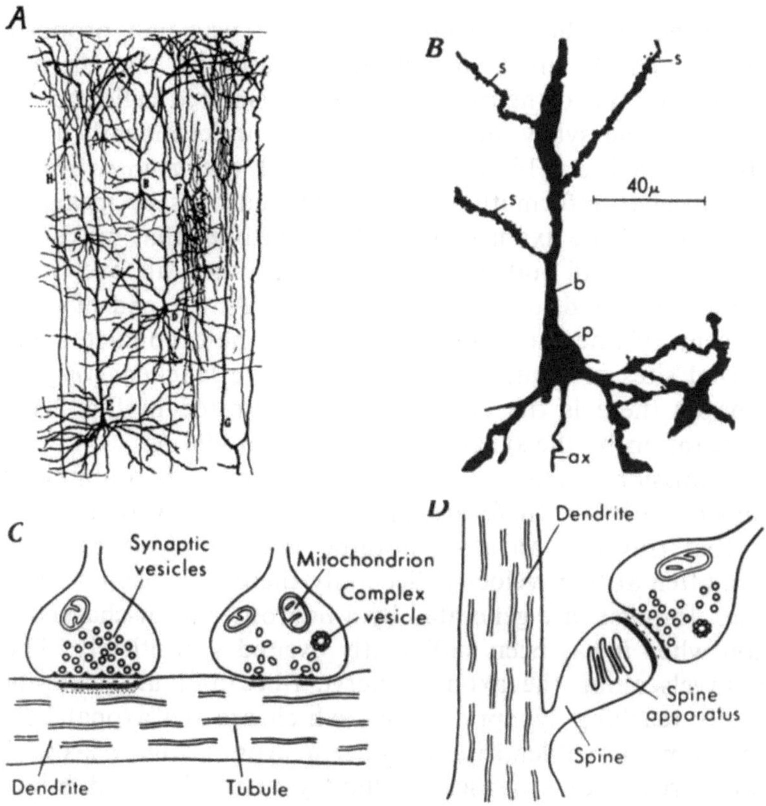

Fig. 2-1 A–D. Neurones and synapses. **A** Pyramidal and stellate cells drawn from Golgi-stained sections of the human frontal cortex. Small (A, B, C) and medium (D, E) pyramidal cells are shown with their branching dendrites and their single axon directed downwards. **B** Golgi preparation of a neurone from cat cerebral cortex with spines (*s*) shown on apical and basal dendrites, but not on the soma (*p*), axon (*ax*) or dendritic stump (*b*). **C** Type-1 (excitatory), type-2 (inhibitory) synapses on a dendrite with the characteristic features displayed diagrammatically. The excitatory synapse has a wider synaptic cleft with a large zone of dense staining. The synaptic vesicles are spherical for the excitatory and elongated for the inhibitory synapse. Special fixation procedures are required for this differentiation. In **D** there is a dendritic spine of a neocortical pyramidal cell with its spine apparatus and an associated type-1 synapse. Whittaker, V.P., Gray, E.G.: The synapse: Biology and Morphology. Br. Med. Bull. *18*, 223–228 (1962)

the dendrites, but not on the soma (p) or the axon (ax). The dendrites are truncated and are seen to be of two varieties, those arising from the apical dendrite (b) of the pyramidal cell, and those directly arising from the soma (p).

At the end of the nineteenth century it was first proposed by Ramón y Cajal, the great Spanish neuroanatomist, that the nervous system is made up of neurones which are isolated cells, not joined together in some syncytium, but each one independently living its own biological life. This concept is called the *neurone theory*. A neurone receives information from other neurones by means of the fine branches of the axons of the other neurones that make contact with its surface and end in little knobs scattered all over its soma and dendrites as indicated in Fig. 2-1 C, D. At the end of the nineteenth century Sherrington proposed that these contact areas are specialized sites of communication, which he labelled *synapses*. At the synapse there is the close contact illustrated in Fig. 2-1 C, D, with separation by the synaptic cleft of about 200 Å.

Transmission in the nervous system occurs by two quite distinct mechanisms. Firstly, there are the brief electrical waves called impulses that travel in an all-or-nothing manner along nerve fibres (axons), often at high velocity. Secondly there is transmission across synapses. Impulses are generated by a neurone and discharged along its axon when it has been sufficiently excited synaptically. The impulse travels along the axon or nerve fibre and all its branches, eventually reaching synaptic knobs, which are the axonal contacts with the somata and dendrites of other neurones. Figure 2-1 C shows the two varieties of synapses, excitatory to the left and inhibitory to the right. The former act by tending to cause the recipient neurone to fire an impulse down its axon, the latter act to inhibit this discharge. There are two kinds of neurones, those whose axons form excitatory synapses and those making inhibitory synapses. There are no ambivalent neurones. Each neurone has thousands of synapses on its surface and it discharges impulses only when synaptic excitation is much stronger than inhibition.

Deep to the cerebral cortex is the white matter that is largely composed of the myelinated nerve fibres, which are the pathways to and from the cerebral cortices. They interconnect each area of the cerebral cortex to lower levels of the central nervous system, or to other areas of the same hemisphere (the association fibres) and of the opposite hemisphere (the commissural fibres). There are about 200 million commissural fibres in the corpus callosum, which is by far the largest system connecting the two hemispheres.

Creutzfeldt (1977) stresses the basic similarity of structure of the whole neocortex, though there are minor differences that form the basis of the subdivisions into the Brodmann areas (Fig. 3-1). Nevertheless the functions of any part of the neocortex are determined by its connectivities in the great system of communication from the thalamic nuclei (cf. Fig. 5-3) to the cortex and by the commissural and association connections of the neocortex.

2.2 Modular Concept of the Neocortex

The great sheet of the neocortex had long ago been divided into various areas on anatomical grounds. The areas defined by Brodmann (1909, 1912) are generally accepted today because they are in good agreement with functionally determined maps. However the subdivision of these large areas was not clearly established until Mountcastle (1957) investigated responses of individual neurones in the somatosensory areas (cf. areas 3, 1, 2 in Fig. 3-1) to inputs from specific peripheral stimulation. He concluded that the data

> support the view that there is an elementary unit of organization in the somatic cortex made up of a vertical group of cells extending through all the cellular layers (cf. Fig. 2-2). The neurons of such a group are related to the same, or nearly the same, peripheral receptive field upon the body surface. They are activated by the same type of peripheral stimulus, i.e. they belong to the same modality subgroup. It follows from this last observation that they should show the same discharge properties, which they do, especially as regards adaptation to steady stimuli. Finally, all the cells of such a vertical column discharge at more or less the same latency to a brief peripheral stimulus.

Thus the long strip of the somatic sensory cortex in Fig. 1-1 has a mosaic-like structure. It will be recognized that the *columnar* or *modular* structure is dependent on the discrete distribution of the thalamo-cortical afferents that convey the sensory input (cf. AFF fibres in Fig. 2-2) (Werner and Whitsel, 1973).

Since that time similar columnar arrangements have been described in the primary visual cortex (Brodmann area 17 in Fig. 3-1), particularly in the exquisitely elaborate investigations of Hubel and Wiesel (1963, 1972, 1974, 1977). The primary auditory cortex (cf. Fig. 1-1) has a similar columnar organization, as shown particularly by Merzenich and Brugge (1973) and their associates Merzenich et al. (1975). There is also a columnar arrangement in the long strip of the motor cortex (cf. Fig. 1-1), as shown by Asanuma and Rosén (1972) and described by Phillips and Porter (1977). Figure 2-3 gives

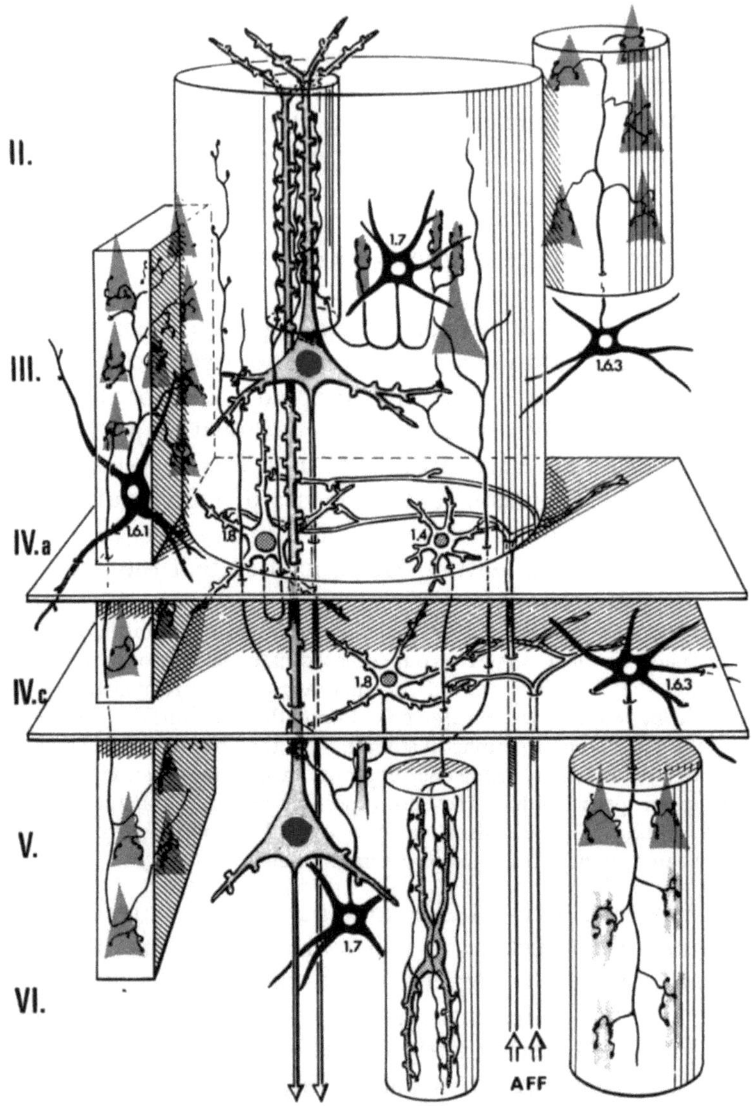

Fig. 2-2. Three dimensional construct showing cortical neurones of various types including two pyramidal cells in laminae III and V with apical dendrites ascending up to lamina I, and axons leaving the cortex. Two specific afferent fibres (*AFF*) are shown branching profusely in lamina IV to make synapses on stellate cells. The cells labelled 1.8 have excitatory axons projecting upwards and making cartridge type synapses (cf. Fig. 7-4) on the apical dendrites of pyramidal cells. Cell 1.4 is a neurogliform cell with descending axon to a Martinotti cell in lamina VI, which has an ascending axon to lamina I as shown in Fig. 7-4. Inhibitory neurones are shown in *black* and may inhibit pyramidal cells in adjacent columns. Further description in text. Szentágothai, J.: Die Neuronenschaltungen der Großhirnrinde. Verh. Anat. Ges. *70*, 187–215 (1976)

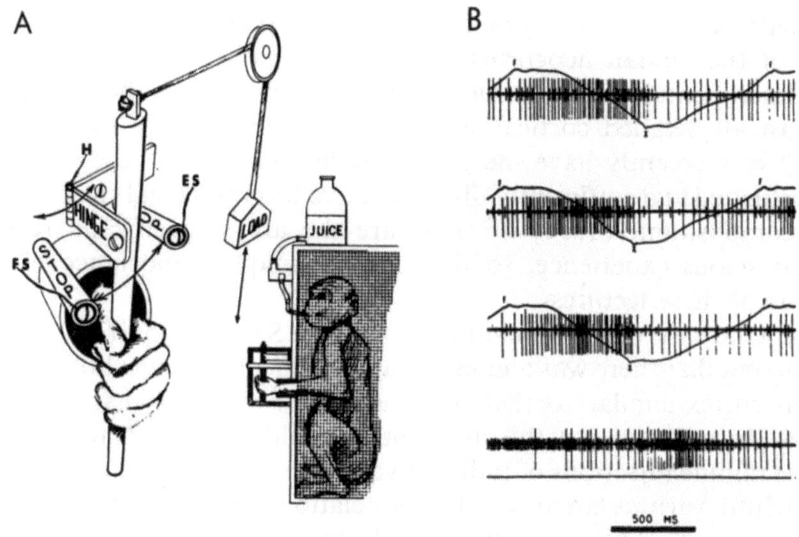

Fig. 2-3 A, B. Pyramidal cell discharges during movements. In an initial operation an electrode was implanted in the monkey's motor cortex in the right place for recording from pyramidal cells concerned in the trained movement, which is shown in detail in **A**, namely the movement of the control bar from one stop to the other. The backward and forward movement is shown in the up and down traces in the three upper frames of **B** along with the discharges of two pyramidal cells, one set large, the other small. This size differential is presumably dependent on the proximity of the cells to the recording electrode. The activities of the two cells were nicely correlated with the downward movement in the three upper frames. However, in the lower frame with some other movement there was no longer correlation. Evarts, E.V.: Unit activity in sleep and wakefulness. In: Neurophysiological Basis of Normal and Abnormal Motor Activity, ed. by M.D. Yahr, D.P. Purpura. Hewlett, New York: Raven Press. 215–253

an excellent example of the firings of two pyramidal cells of the motor cortex that are in relation to the arm movements.

An excellent survey of all this work on the primary sensory and motor cortices has been given by Mountcastle (1978 b), and the modular composition of the visual cortex has been illustrated by figures from Hubel and Wiesel in the previous lecture series (Eccles, 1979 a, Figs. 8-12 and 8-13). It is generally agreed that the primary cortical areas are not directly concerned with conscious perception or with the initiation of voluntary actions. Moreover, as will later appear, the columnar organization has a different structural basis from that of the remainder of the cortex. It is important to realize that these primary sensory and motor areas together make up no more than 5% of the human neocortex. However, because of their openness to experimental investigation by selective inputs from the periphery, almost all detailed cortical studies have been made in these areas. Only very recently have the great association areas (the 95%) been investigated at a sufficiently discriminative level to reveal their modular composition. Moreover these are the areas specifically related to conscious experience, so they are of unique importance to the theme of these lectures.

On the basis of the similar microstructure Szentágothai (1972, 1975) suggested that there was a modular arrangement for the whole association cortex similar to that demonstrated for the primary sensory areas. In order to test this assumption Goldman and Nauta (1977) made minute injections of radioactive tracers ([^{3}H]leucine or [^{3}H]proline) into various areas of the association cortex of monkeys. It is well known that these amino acids are taken up by the neurones and converted into labelled proteins which are then transported along the axons of the neurones to their terminal branches. It is not necessary to go into the details of this reliable technical procedure of radio-labelling for determining the distribution of the pyramidal cell axons of a small cortical area. As recognized by Szentágothai (1978a), the radio-labelling revealed a quite unexpected mode of distribution of the axons that is of the greatest importance.

Figure 2-4A shows an autoradiogram of columns in the neocortex far removed from the site of the cortical injection in the frontal lobe. There are three well defined columns with different degrees of radio-labelling and one very faint column to the right. Goldman and Nauta give the dimensions as 200 to 400 µm across and point out that the labelling extends perpendicularly across all the layers of the cortex, being especially prominent in lamina I (cf. Fig. 2-6). Note the dark spaces between with a grain density hardly above background. It is important to recognize that the labelling is in the association fibres that have come from the injected area to terminate in these labelled columns. In Fig. 2-4B the sectioning of the cortex was oblique with the consequence that the columns were cut across, there being four nicely separated ovoid labelled areas about 300 to 500 µm across. In interpreting these results account must be taken of the size of the area injected with the radiotracer, which would include many modules. Thus the observed projections in Fig. 2-4 would be from an assemblage of modules. These observed

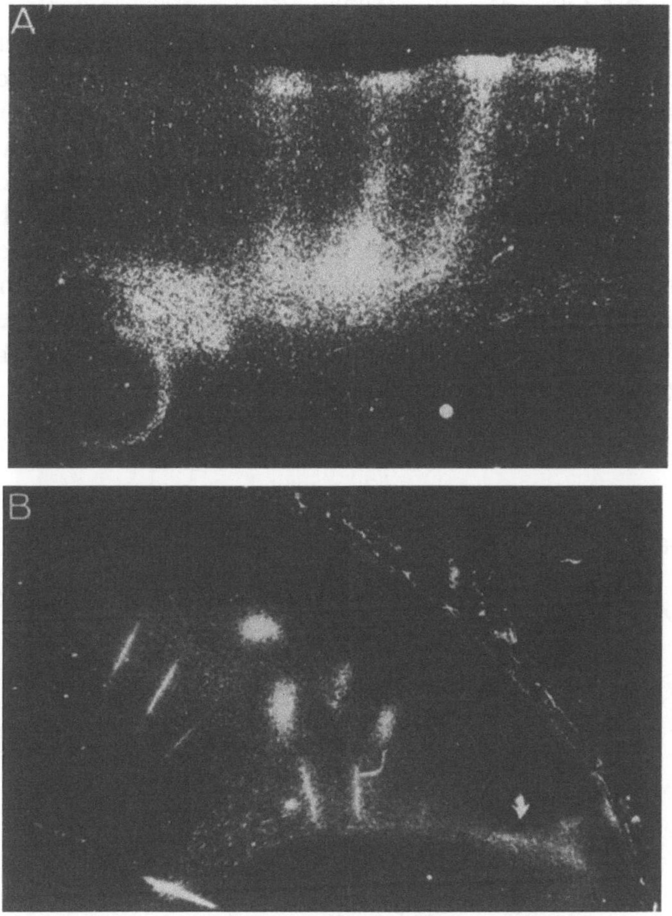

Fig. 2-4. A Dark-field autoradiogram illustrating columns in the ipsilateral retrosplenial cortex following a single injection of [^{3}H]leucine and [^{3}H]proline into the dorsal bank cortex of the principal sulcus in a 4-day-old monkey. Cingulum bundle, containing labelled axons, is to the *lower left* of center. ×20. **B** Dark-field autoradiogram illustrating columns in the contralateral homotopical cortex from the injection site in the hand and arm area of a 4-day-old monkey. *Arrow* indicates single band of grains distributed over all layers of cortex at ventral boundary of the projection field. ×10. Goldman, P.S., Nauta, W.J.H.: Columnar distribution of cortico-cortical fibers in the frontal association, limbic and motor cortex of the developing rhesus monkey. Brain Res. *122*, 393–413 (1977)

projections are very widely distributed over the ipsilateral cortex, the association connections; and also they are distributed to the contralateral cortex, the commissural connections that are mostly between mirror-image areas, but not entirely so.

In Fig. 2-5 A Szentágothai (1978 a) gives a diagrammatic representation of this extraordinary finding that the immense sheet of the neocortex is subdivided into a mosaic of quasi-discrete space units. The thesis will be developed that these space units are the modules which form the basic anatomical elements in the functional design of the neocortex. Twelve closely packed pyramidal cells are shown in one such module or column in the left of Fig. 2-5 A, which has a width of about 250 μm. The axons from these pyramidal cells project to three other modules of that same hemisphere, and, after traversing the corpus callosum, to two modules of the other hemisphere. Thus we have simply displayed the association and the callosal projections of the pyramidal cells of a module.

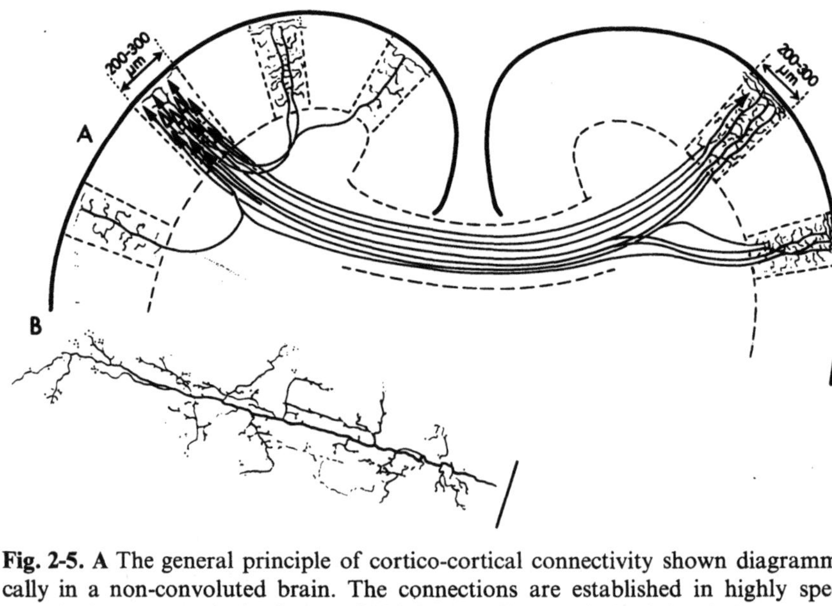

Fig. 2-5. A The general principle of cortico-cortical connectivity shown diagrammatically in a non-convoluted brain. The connections are established in highly specific patterns between vertical columns of 200–300 μm diameter in both hemispheres. Ipsilateral connections are derived mainly from cells located in layer III (cells shown at left in *outlines*), while contralateral connections (cells shown in *full black*) derive from all layers II–VI. The diagram does not try to show the convergence from afferents originating from different parts of the cortex to the same columns. **B** Golgi stained branching of a single cortico-cortical afferent, oriented in relationship to the module with a single afferent in **A**, but at several times higher magnification. It illustrates the profuse branching in all laminae. Bar = 100 μm. Szentágothai, J.: The neuron network of the cerebral cortex: A functional interpretation. Proc. R. Soc. Lond. B *201*, 219–248 (1978)

We should notice several important features of Fig. 2-5 A. Firstly, several pyramidal cells of a module project in a *completely overlapping manner* to other modules which are so defined and which are again about 250 μm across. In Fig. 2-5 A this dimension is already represented by the branches of a single association fibre for two modules and the overlapping distribution is illustrated for two association fibres and for four and five callosal fibres. Secondly, the callosal projection is mostly but not entirely to symmetrical modules on

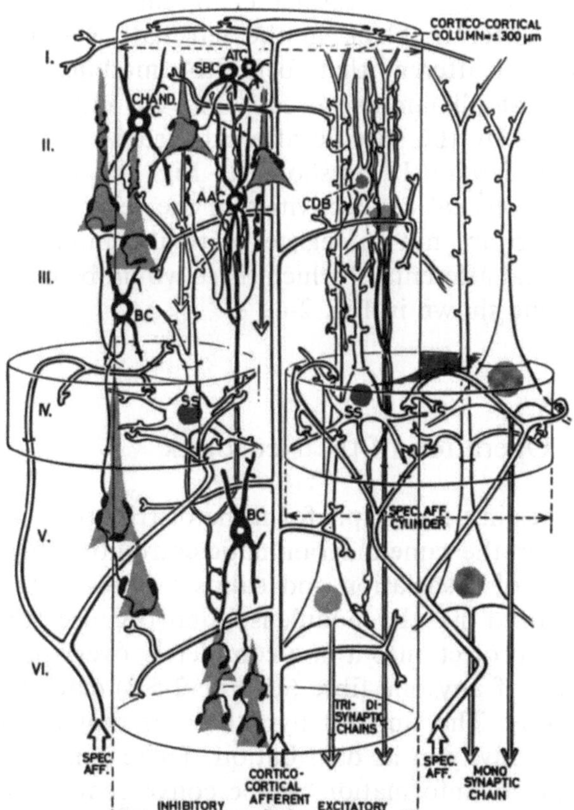

Fig. 2-6. Diagram illustrating a single cortico-cortical column and two specific subcortical afferent arborization cylinders. Lamination is indicated on the left margin. The *right* half of the diagram indicates impulse processing over excitatory neurone chains, while the *left* half shows various types of inhibitory interneurones in *full black*. Further explanation in the text. Szentágothai, J.: The local neuronal apparatus of the cerebral cortex. In: Cerebral correlates of conscious experience. Buser, P., Rouguel-Buser, A. (eds.), 131–138 Amsterdam: Elsevier 1978

the contralateral side. Thirdly, there is reciprocity of callosal connection between symmetrical modules. Below Fig. 2-5A there is shown in B at higher magnification the faint arborization of a single cortico-cortical afferent fibre, as revealed by Golgi staining. It is seen to traverse all laminae of the cortex and to branch extensively so that it is distributed to a column about 200 μm across, as is drawn in Fig. 2-5A.

Figure 2-6 provides a diagrammatic summary of the essential components of the neocortical module together with the excitatory and inhibitory neurones, the former being shown in outline in the right half, the latter in black in the left half. Centrally there is the cortico-cortical afferent that forms the module (cf. Fig. 2-5B) by its branches at all laminae except IV. On either side are the two discs formed by the specific afferents from the thalamus with their branches in lamina IV as shown in Fig. 2-2. From these discs there extend the vertical connectivities of the relay neurones shown in Fig. 2-2, which are not in register with the module defined by the cortico-cortical afferents of which there would be many hundreds instead of the one shown in Fig. 2-6.

2.3 Modular Operation of the Neocortex

Figure 2-5A gives in very simple form the distribution of the output from a module. In the same diagrammatic convention Fig. 2-7 shows the convergence of association and callosal fibres onto a module. It can be recognized that the module is a unit because other modules of the neocortex project onto it in a completely overlapping manner, for the branches of any one fibre (cf. Fig. 2-5B) extend throughout the whole module. The simplest hypothesis is that this convergent input has no selectivity in its distribution. There is merely a pooling of the total input of information by the convergent association and callosal fibres. The complex integrative machinery (cf. Fig. 2-6) of the whole module then operates to determine the output from the module from moment to moment. At any one instant the output will have an intensity compounded of the number of pyramidal cells firing and the frequencies of these firings. It will also have a temporal pattern given by the time course of the integrated frequency. An additional action of the module is to inhibit adjacent modules (cf.

Fig. 2-2) by the lateral distribution of inhibitory neurones (Marin-Padilla, 1970), and reciprocally it will receive inhibition from these surround modules.

In order to gain some insight into the manner of operation of this neuronal machine, we have to recognize that Fig. 2-5A is greatly simplified. In the first place there are about 2500 neurones in a module 250 μm across, at least 1000 of which are pyramidal cells. On the basis of the fibre count for the corpus callosum (200 million), there would be about 50 pyramidal cells projecting from one module to make callosal afferents to the other side, instead of the six in Fig. 2-5A. On an estimate of ten times as many association fibres than callosal fibres, there would be 500 pyramidal cells as association projectors instead of the three in Fig. 2-5A. Thus there would be a much more intense convergence of inputs from any one module than is represented in Fig. 2-5A, and also probably dispersion to many more modules. The projection from module to module is not random. On the contrary it is very highly structured in accord with specific distributions, as is indicated for the sequential relays of the somatosensory and visual systems in Fig. 3-2, where there is also shown convergence between these large Brodmann areas. We can think of the modular connectivities of Figs. 2-5 and 2-7 as giving the grain of the large association connectivities. Szentágothai (1978a) suggests convergent and divergent numbers as high as 50.

But, even if this massive addition were made to Fig. 2-5A, the diagram would give the situation of connectivities for only a moment of time. If we assume a strong excitation by the convergent fibres onto the primary module in Fig. 2-7 with a powerful burst of impulse discharges from the constituent pyramidal cells, there will be within a few milliseconds strong excitatory inputs into many other modules by the association and callosal projections, as shown in Fig. 2-5A. Some of these secondary modules in turn will be excited sufficiently to discharge to tertiary modules, and these in turn to quaternary. Some idea of the fast transmission from module to module is given by the observation of Newman (1976) that in response to a brief click some neurones of the dorsolateral prefrontal cortex respond with a latency as brief as 20 ms. At least two cortical relays by modular responses would be involved in this brief latency, as well as the transmission time to the primary auditory cortex. In a small fraction of a second we have in effect a spreading pattern of excitation that is not random, but strictly specified by the sequential projections

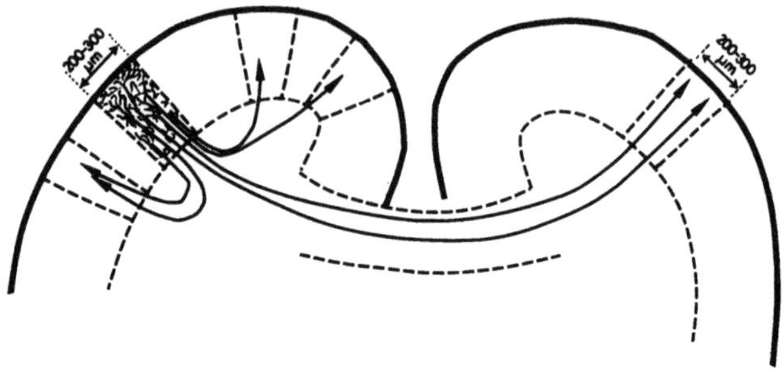

Fig. 2-7. Plotting of modular connectivities by association and callosal fibres as in Fig. 2-5A, but in the reverse direction in order to show convergence of projections from several modules onto one in a completely overlapping manner as described in the text

from the secondary, tertiary, etc. modules in this particular instance (cf. Fig. 2-8). There are approximately 4 million modules in the human neocortex, so there are immense possibilities for developing spatio-temporal patterns even on the simple assumption that each module operates as a unit in its reception and projection. However, there is certainly a gradation of the responses of the pyramidal cells in a module, which is exhibited by the wide range in their firing frequencies, as reported by many investigators.

The special design features of a cortical module are: firstly, the internal pattern of connectivities of its constituent neurones (Fig. 2-6); secondly, the inhibitory surround (Fig. 2-2) that is built up by some of its constituent inhibitory neurones and that sharpens its boundaries (Szentágothai 1978a); thirdly, the output, which is graded both by number of pyramidal cells that are excited to fire impulses and by the frequencies of this firing (Fig. 2-3). The simplest hypothesis is that a module acts as a fundamental unit in the performance of the neocortex and that fractionation of a module has no functional significance because there is complete overlap of the projections of its constituent pyramidal cells (cf. Figs. 2-5 and 2-7). Each module acts as a unit processing the many convergent inputs from the other modules and in turn it projects divergently to many other modules. It thus embodies the Sherringtonian principles of convergence and divergence. As shown in Fig. 2-9 even with minimal connectivities, a module (with star) can participate in two quite distinct spatio-temporal patterns. With the modular connectivities of the neocortex

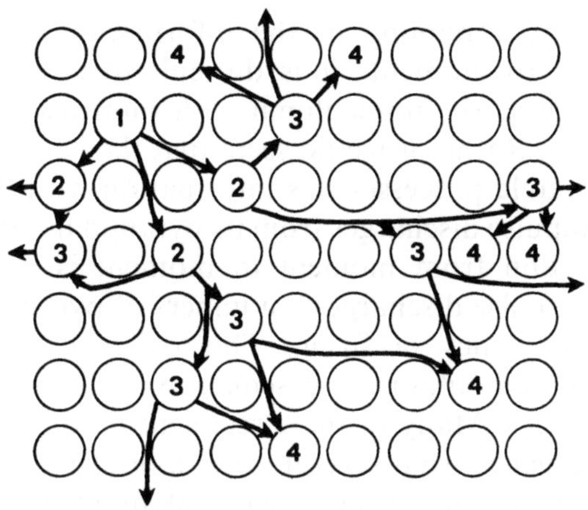

Fig. 2-8. Diagrammatic representation of cortical modules as seen from above, i.e. each is seen as its projection to the surface about 250 μm across. In order to allow drawing of transmission pathways from module to module, the modules are drawn well separated and not in the actual close juxtaposition. Strong activation of module *1* as in Fig. 2-5 leads to transmission effectively exciting modules labelled *2*, and these in turn to the modules (*3*) and then modules (*4*). Further description in text

a module would be able to participate in a prodigious number of spatio-temporal patterns.

In contrast to this simple modular hypothesis there has been speculation concerning the possible subdivision of the basic modules that are depicted in Figs. 2-5 and 2-7. For example, Mountcastle (1978 b) postulates a much smaller unit, a minicolumn of about 30 μm across, and Szentágothai (1978 a) suggests that there is an ultimate refinement of a module by interaction of its internal excitatory and inhibitory interneurones to give a high degree of individuality in output, even to a single pyramidal cell. I prefer to consider the whole module as a processing unit with many output lines in parallel, so that there will be effective synaptic excitation through convergence on the modules to which it projects (cf. Figs. 2-5 and 2-7).

It might be considered that the module is too crude a unit to handle and store the immense quantity of information that is the requirement for a lifetime of operation of the human brain. Such a consideration gives the attraction to the mini-column hypothesis. Nevertheless there are two counter-arguments.

In the first place, because of the completely overlapping distribution of the association and callosal inputs into a module (Figs. 2-5 and 2-7), there is a pooling of the information converging onto it from moment to moment. It is difficult to see how a fraction of a module such as the minicolumn of Mountcastle can have any selective function. Rather it would seem that the neuronal machinery of the whole module processes this information with the consequent generation of impulse discharges from its pyramidal cells. The intensity of the excitation from moment to moment is signalled by the integral of the impulse discharges – numbers of pyramidal cells times their average frequency of firing. This episodic output signal is distributed to the modules next in sequence in the divergent manner indicated in Fig. 2-5. Many of these recipient modules will of course be foci of convergence from several other modules that are activated in the same episode, and hence each will momentarily experience an intensity of excitation that can have a wide range of variance and so exhibit a corresponding variance in output discharge, and so on sequentially.

In the second place, it is evident that modules are very rich in their pattern-generating capability even if they operate globally without subdivision into minicolumns (cf. Fig. 2-9). At an approximate estimate there could be in the human neocortex 4 million modules that generate the spatio-temporal patterns which encode the total cognitive performance of the human brain – all sensing, all memories, all linguistic expression, all creativities, etc. We may ask: is 4 million adequate for this tremendous task? The only answer I can give is to refer to the immense potentialities of the 88 keys of a piano. Think of the creative performances of the great composers, Mozart, Beethoven and Chopin for example. They could

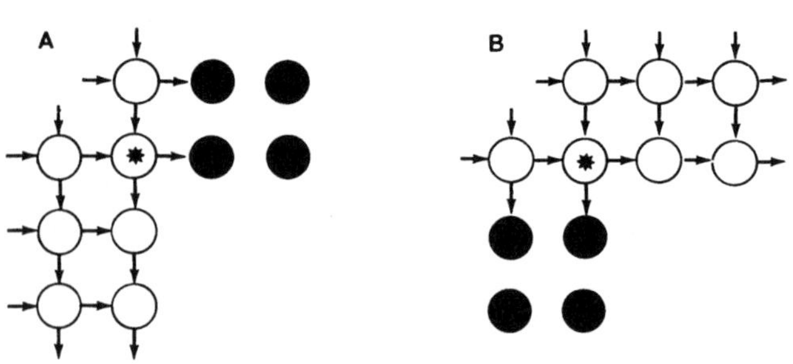

Fig. 2-9. Simplest diagram to show that a module (*starred*) can participate in two quite distinct spatio-temporal patterns, *A* or *B*, in accord with two different ongoing activations of other modules. It is assumed that convergence from two modules is necessary for evoking a discharge of a module in an all-or-nothing manner

utilize only four parameters in their creation of piano music with the 88 keys, each of which has an invariant pitch and tonal quality. And a comparable four parameters are utilized in creating the spatio-temporal patterns of activity in the four million modules of the human neocortex. I will consider these four parameters in turn, pointing out the comparable features.

Firstly, there is intensity. In music this is the loudness of the note, whereas with the module it is the integral of the impulse firing in the output lines from the module as described above. It has to be recognized that in the analogy the keys of the piano have a pitch differentiation, whereas in the neocortex, each module (or key) has its unique connectivity (cf. Fig. 3-2) which is the basis of the coded properties that give the modality of the perception read out from it, e.g. light, or colour or sound, as well as all the subtle perceptual qualities.

Secondly, there is the duration of the note or of the impulse firing from the module.

Thirdly, there is the sequence of the notes which gives a melody with piano music and the temporal pattern of modular activities in the neocortex. One can imagine that this modular pattern is played as a sequential transmission of modules to modules to modules (cf. Fig. 2-8).

Fourthly, there is the simultaneous activation of several notes, a chord, or of several modules in the brain. With the chord the maximum number is ten, with the modules it may be thousands.

It will I think be recognized that the enormous generation of musical patterns using the 88 keys of a piano points to a virtually infinite capacity of the 4 million modules to generate spatio-temporal patterns. In Fig. 1-7 the modules are shown as 40 vertical bands. If this number were increased 50,000 times, it would approximately correspond to the number of modules that potentially are capable of contributing to the conscious experiences that can be derived from the brain. However it must be realized that, just as with piano keys, the experiences are dependent on the play of the four parameters listed above.

We can imagine that the intensities of activation of modules are signalled symbolically by the lighting up of modules. Using this analogy for nerve cells Sherrington (1940) gave a wonderful picture of the patterned lights of the active brain. So if we could see the modules lighted up on the surface of the neocortex, it would present an illuminated pattern of 50 cm by 50 cm composed at any moment of modules 0.25 mm across that have all ranges from dark to dim to brighter to brilliant. And this pattern would be changing in a scintillating manner from moment to moment giving a sparkling pattern woven by the 4 million modules. An apt analogy is provided by a TV screen where there are up to 1 million units composing the picture. This symbolism of patterned lights gives some idea of the immense task confronting the mind in deriving conscious experiences from the neocortical modules. One suspects that the dark or dim modules are usually ignored, and that attention is rivetted by the brilliant – or rather by the patterns of brilliance.

2.4 Liaison Between Brain and Mind

So far the description of the modular operation has been at a purely materialistic level. For example, it could be an account of the operations of the neuronal machinery from sensory input through immensely complicated spatio-temporal patterns of modules eventually to some motor output. But on the dualist-interactionist hypothesis we have to raise the question: what activity of the neuronal machinery could be 'read out' by the self-conscious mind? We can firstly say that it is inconceivable that the self-conscious mind is in liaison with single nerve cells or single nerve fibres. These neuronal units as individuals are far too unreliable and ineffective. In our present understanding of the mode of operation of neuronal machinery, emphasis should be placed on ensembles of neurones (hundreds or thousands) acting in some collusive patterned array. Only in such assemblages can there be reliability and effectiveness. The modules of the cerebral cortex are such ensembles of neurones. The module has to some degree a collective life of its own with about 2500 neurones of diverse types and with a functional arrangement of feedforward and feedback excitation and inhibition (Szentágothai, 1978a, 1978b, 1979). As yet we have little knowledge of the inner dynamic life of a module. I know of only one thorough systematic study of the individual neurones in a module with the derivative of a functional diagram of their connectivities. It was in the primary visual cortex (Toyama et al., 1974, 1977a, 1977b). We may conjecture that, with its complexly organized and intensely active properties, a module could be a component of the physical World (World 1) that is open to the self-conscious mind (World 2) both for receiving from and for giving to. In Fig. 1-7 the modules are indicated as vertical bands across the whole thickness of the cerebral cortex. It can further be proposed that not all modules in the cerebral cortex have this transcendent property of being 'open' to World 2, and thus being the World 1 components of the interface. By definition there would be restriction to the modules of the liaison brain, and only then when they are in the correct level of activity.

Figure 2-10 gives a diagrammatic illustration of the conjectured relationship of open and closed modules as viewed by looking down at the surface of the cortex. A convenient diagrammatic liberty is to show the columns as separate discs, and not in the close contiguity that is the actual relationship (Szentágothai, 1978a). Furthermore, it has to be recognized that the normal intensely dynamic situation is frozen in time. The convention is that open modules are shown as open circles, closed as solid and that there are also partly open modules. It can be conjectured that the self-conscious mind probes into this modular array, being able to receive from and give to only those modules that have some degree of openness. However,

Fig. 2-10. Diagrammatic plan of cortical modules as seen from the surface (cf. Fig. 2-8). As described in the text the modules are shown as circles of three kinds, *open, closed (solid black)* and *half open*. Further description in text

by its action on open modules, it can influence closed modules by means of impulse discharges along the association fibres from the open modules, as already described (Fig. 2-5 and 2-7), and may in this manner cause the opening of closed modules. Figure 2-10 has been designed to show several nuclei of excited modules interspersed in large dark areas. Also there are smaller patches of excitation. Figure 2-10 represents just a fragment of a coded representation in a moment of time. The complexity of the real situation can be appreciated when it is recognized that the modular assemblage of the neocortex would be represented by increasing the dimensions of Fig. 2-10, 80-fold in each direction.

The simplest hypothesis of brain–mind interaction is that the self-conscious mind can scan the activity of each module of the liaison brain or at least those modules tuned in to its present interest. It has already been conjectured that the self-conscious mind has the function of integrating its selections from the immense patterned input it receives from the liaison brain – the modular activities in this present hypothesis – in order to build its experiences from moment to moment. The modules selected in this way constitute for the moment the World 1 side of the interface between World 1 and World 2, as diagrammatically shown in Fig. 1-7. This interface is thus a constantly changing territory within the extensive area of the liaison brain.

Experimental evidence indicates that a large patterned complexity of modular operation must be built up before there is a conscious experience. There is for example a delay of up to 0.5 s before weak repetitive electrical stimulation of the somaesthetic cortex gives rise to a conscious experience of touch (Lecture 3; Libet, 1973). As will be fully described in Lecture 4, Desmedt and Robertson (1977a, 1977b) have shown that, with weak electrical stimulation of the 'attending' finger, conscious recognition of the stimulation is associated with delayed negative and positive potentials (for up to 0.5 s) widely dispersed over both hemispheres, there being no such late potentials with similar stimulation of fingers not being attended to. Conversely 'willing' a movement is associated with a widely dispersed potential (the readiness potential) that builds up during 0.8 s before the onset of a simple movement and that presumably is generated by immensely complicated patterns of modular operation (Lecture 4, Fig. 4-9; Kornhuber, 1974; Popper and Eccles, 1977, Chap. E3).

It appears that in the primate the pattern of modular projection to other modules is established at the latest within a few days after birth (Goldman and Nauta, 1977). If the whole modular connectivity is thus prewired before use, it might appear difficult to account for the changed performance of the neocortex that presumably is the neural basis of learning. Brief reference should therefore be made to an hypothesis of cognitive learning (Lecture 7 and Eccles, 1979a, Lecture 9) in which it is proposed that in learning there is an hypertrophy of synapses made by horizontal fibres on the apical dendrites of pyramidal cells in the most superficial laminae of the cerebral cortex (laminae I and II). It is postulated that these synapses are caused to hypertrophy in a specific pattern because of the selection by conjunction in time of a specific input from the hippocampus on the one hand and a cortico-cortical input on the other (cf. Figs. 7-5, 7-6). In this way a selection of adjacent modules would come to be more effectively linked together, (cf. Fig. 7-7), and hence there would be a change in the evolving modular pattern that could be the basis of the learned performance of the neuronal machine. In the radio-tracer observations of Goldman and Nauta (1977) the radioactive modules always have an increased intensity in laminae I and II (Fig. 2-4A) that could be evidence for such synaptic hypertrophy. It would be of great interest if the intense radioactivity found by Goldman and Nauta (1977) in laminae I and II would extend in a patchy manner from the radioactive module.

We have to consider the arrangements for modular interaction through pyramidal cells of other modules. Thus each module is projecting to many others and they in turn to many others. So there are long and complex patterns of this mutual interaction. I conjecture that the self-conscious mind acts in modifying slightly some of these modules, presumably hundreds, and that the modules collectively react to these modifications which are transmitted by the circuits of the association fibres and the callosal fibres. In addition the self-conscious mind is all the time reading out or perceiving the responses that it is making in this subtle manner. It is an essential feature of the hypothesis that the relations between modules and the self-conscious mind are reciprocal, the self-conscious mind being both an activator and a receiver, as is indicated by the reciprocal arrows across the frontier in Fig. 1-7. This reciprocity of action conforms with Einstein's statement that 'it is contrary to the mode of thinking in science to conceive of a thing ... which acts of itself, but which cannot be acted upon'.

In further developing the hypotheses of some modules being open to World 2 in the guise of the self-conscious mind, it may be supposed that the self-conscious mind does not make a superficial pass over the module, as may be imagined if it merely sensed the micropotential fields in the areas (cf. Pribram, 1971). Rather it may be envisaged that it 'probes' into the module, reading out and influencing the dynamic patterns of neuronal performance. It can be assumed that this is done from moment to moment over the whole scattered assemblage of those modules processing information of immediate interest (attention) to the self-conscious mind for its integrational performance.

Kornhuber (1978) rightly stresses the importance of order in the natural world, which has two components – order and energy. Information is an aspect of order. The general thesis is that this growth of order in the biological evolution of the brain resulted in the emergence of consciousness. Such concepts also help in the interpretation of the mode of transmission across the frontier of Fig. 1-7. It could be transmission of order or information not of energy, as has already been noted in Lecture 1. By its very nature the World 2 of conscious experiences must be replete with order and information.

When we consider the immensely patterned structure of modules, the minute cortical area of Fig. 2-10 being extended about 80 times in each direction, it is difficult to understand how the self-conscious mind can relate to such an enormous complexity of spatio-temporal patterns. This difficulty is mitigated by three considerations. Firstly, we can realize that our self-conscious mind has been learning to accomplish such tasks from our babyhood onwards, a process that colloquially is called 'learning to use one's brain'. Secondly, by the process of attention (cf. Lecture 4) the self-conscious mind selects from the total ensemble of modular patterns those features that are in accord with its present interests. Thirdly, the self-conscious mind is engaged in extracting 'meaning' from all that it reads out. This is well illustrated by its interpretation of the many ambiguous and impossible figures that will be considered in Lecture 3.

There are some parallels between the modular hypothesis, as here described, and the suggestions of Mountcastle (1978b) that modules linked together in echeloned parallel and serial arrangements compose 'distributed systems'. These systems link internally generated neural activity with the input–output function of the nervous system. It is proposed that they 'provide an objective mechanism of conscious awareness' in a manner exactly consonant with that of Psychoneural Identity. Similarly Szentágothai (1979) concludes that 'dynamic patterns' might 'give a scientific explanation of the higher functions of the brain, including even consciousness'. Edelman (1978) likewise suggests, on the basis of a model derived from immunological theory, that

> the brain processes sensory signals and its own stored information upon this selective base in a phasic (cyclic) and re-entrant manner that is capable of generating the necessary conditions for conscious states.

These hypotheses are essentially versions of the psychoneural identity explanation of consciousness. These recent attempts at formulating a solution of the brain–mind problem that is consonant with the closedness of World 1 seem to be no more than vaguely expressed examples of the promissory materialism criticized by Popper and Eccles (1977, pp. 96–97). Moreover these attempts can be subjected to the same criticisms that have been raised in Lecture 1 against less sophisticated forms of the psychoneural identity theory. These theories are all confronted by the impossible task of deriving a mental world out of a material world of neuronal circuits. By contrast dualist-interactionism postulates the existence of both a mental world and a material world, and is concerned essentially with their interaction.

Granit (1976) and Thorpe (1978) have recently made interesting and important contributions to the role of 'purpose' in the behaviour of higher animals. There seems to be no doubt that higher vertebrates, birds and mammals, exhibit a true purposive or insightful behaviour that leads to the supposition that they have conscious experiences (Lecture 8). But presumably such consciousness would differ greatly from human self-consciousness, as has been argued by Popper and Eccles (1977, Dialogue, II).

2.5 Summary and Conclusions on the Brain–Mind Problem

A brief outline of the hypothesis may be given as follows. The self-conscious mind is actively engaged in reading out from the multitude of liaison modules that are largely in the dominant cerebral hemisphere. The self-conscious mind selects from these modules according to attention and interest, and from moment to moment integrates its selection to give unity even to the most transient experiences. Furthermore the self-conscious mind acts upon these modules modifying their dynamic spatio-temporal patterns. Thus it is proposed that the self-conscious mind exercises a superior interpretative and controlling role. A key component of the hypothesis is that the unity of conscious experience is provided by the self-conscious mind and not by the neural machinery of the liaison areas of the cerebral hemisphere. Hitherto it has been impossible to develop any neurophysiological theory that explains how a diversity of brain events comes to be synthesized so that there is a unified conscious experience of a global or gestalt character. The brain events remain disparate, being essentially the individual actions of countless neurones that are built into modules and so participate in the spatio-temporal patterns of activity. The brain events provide no explanation of our commonest experience, the visual world seen as a global entity from moment to moment.

The philosophy of a strong dualist-interactionism as developed in this lecture has led me to build up conjectures upon the most advanced concepts at present available on the structure and functioning of the neocortex. It must be realized that not until 1977, did we have some knowledge of the basic unit of operation of the association neocortex, the module. I have endeavoured to keep the story as simple as possible. Yet how should we expect to have an easy solution of this greatest problem confronting us, namely the brain–mind problem? I do not claim to have offered a solution, but rather to have indicated in general outline the way the solution may come. In some mysterious manner the human brain evolved with properties of a quite other order from anything else in nature.

It is a measure of our ignorance that in the neocortex no special structural or physiological properties have been identified that distinguish sharply a human brain from the brain of an anthropoid ape. The tremendous difference in performance can hardly be attributable to a mere threefold increase in modules. The hope is that we may be challenged to enter on a new scientific era where the structure and function of the human neocortex will be intensively studied by the most advanced techniques enlightened by a superb creative imagination (cf. Lecture 7).

Lecture 3

Sensory Perception and Illusions

Résumé

The nature of sensory perception is discussed in philosophical and psychological terms with special reference to meaning and action.

There is an account of the connectivities from primary to secondary to tertiary cortical sensory areas. Particular reference is made to communication by the cortical modules in the transitional zones from the primary sensory areas to the general association areas.

The principal part of the lecture is concerned with visual illusions. The physical and physiological types are mentioned, but attention is concentrated on the various kinds of *cognitive illusions* because they provide important tests of the brain–mind theories. Many of these illusions are described and some are illustrated.

It is claimed that with respect to cognitive illusions the dualist-interactionist hypothesis has a greater explanatory power than the materialist theories of the mind. It is argued that cognitive illusions arise because the self-conscious mind strives for a meaningful interpretation of the on-going cerebral activity.

Finally there is consideration of a related type of brain–mind-interaction that occurs in *depth perception,* which is an experience derived from many features of the visual input. Again it will be shown that dualist-interactionism can provide a general explanation of the experience of depth perception, which has not so far been possible on even the most sophisticated materialist theories. At the best these theories provide models that could simulate illusions.

3.1 Introduction

Sensory perceptions are conscious experiences deriving from inputs to the brain from sense organs. Such perceptions can be synthesized from an immense variety of inputs from any one sensory system, say vision, but also there can be a synthesis of inputs from two or more different sensory systems, as for example when we palpate an object that we see, the better to objectify it. It must be recognized that perceptions are subject to the limitations imposed by the sensory systems. Vision and hearing have sharply restricted ranges of spectral sensitivity. Many substances are odourless and tasteless. It must further be recognized that our perceptions give us symbolic information about the external world, there being quite extraordinary transmutations: electromagnetic radiations with wave lengths from 380 to 720 nm are perceived as coloured lights in the spectral range from violet to red; pressure waves in the atmosphere with frequencies from 15 to 20,000 Hz are perceived as sounds from an extremely low to an extremely high pitch; different chemical substances evoke a fascinating range of olfactory experience.

However these crude perceptions give no insight into the extraordinary problems raised in trying to account for the rich experiential imagery that is given to us from moment to moment in our waking life. We think we perceive a real external world, including our own bodies, as it actually is with its form, its spatial relations, its colours, its sounds, its material structure, as well stated by Held (1965):

> Ordinary experience of objects seems to raise no problems. The observer opens his eyes and sees his environment; he shuts them and blots out the view. He walks among objects and gains different perspectives of them. He finds that he can touch most objects that are within reach. He talks to others about these objects and shares expectations concerning them. Such common observations convince us that the familiar world of objects has a continuity in time and space which is independent of our scrutiny. The eyes provide vignettes of this world; different senses make other properties accessible. There seems to be nothing problematic in this description of the observer; one would expect that untutored men throughout the ages have shared it. How then does a need for explanation arise?

It arises because severe problems develop from our modern scientific knowledge of the phenomena concerned in perception, both in the world of physics and in the structure and functioning of our sensory systems both peripherally and centrally in the brain.

Jung (1980) stresses the importance of meaning and action in perception:

> Physiological as well as psychological experience demonstrate two essential integrative effects of sensory perception:
> 1. Sensory stimuli are effective in animals and man mainly for their *meaning* and less by their intensity, quality or modality:
> 2. The organism learns from perceptions mainly *during action* on the basis of emotional and instinctive motivations. Both involve an anticipation of the sensory stimulus.

Furthermore Jung (1954) proposed a model of the selective function of consciousness and attention as a searchlight, proposing that (Jung, 1978):

> in the theatre of consciousness the various actors emerging on the stage from unseen back-stage areas represent the contents of conscious experience evoked from the unconscious. The backstage corresponds to the memory stores of the inner world and to the sensory stimuli of the outer world. Both are selected and evoked by attention to appear in conscious experience. This process is directed by the stage manager, representing the will and the emotions, who co-ordinates the action and the illumination of the actors and stage by the attentive searchlight, just as we select percepts and memory recollections in willed and emotional activity.

Throughout all this quotation there is a clear dualistic mode of thought and expression. We will be returning to these ideas in Lectures 4 and 5, but they are of particular importance in relation to illusions that will form the main theme of this lecture.

3.2 Anatomical Considerations (Jones and Powell, 1973)

It is generally agreed that perceptual experiences are not generated when the inputs from receptor organs activate neurones in the primary sensory areas, as is indicated by the arrows from eyes, ears and body in Fig. 1-1. In those primary areas there is processing of the input information through relays to interneurones and eventually to the pyramidal cells, as is indicated in the greatly simplified diagram of a column or module in Fig. 2-2. The afferent fibres (AFF) from the thalamic nuclei relaying the sensory information branch profusely in lamina IV. There they relay synaptically to excitatory neurones that are vertically oriented to build up a powerful excitation that is exerted onto the pyramidal cells (two being shown in Fig. 2-2) which project out from the column. There is also shown in Fig. 2-2 the inhibitory neurones (indicated in black) that mostly act on adjacent columns, as already described in Lecture 2. The projections by the pyramidal cell axons to other cortical areas represent the transition from the columns structured by the thalamic inputs (Fig. 2-2) to the association connectivities that structure the modules (cf. Fig. 2-6) for the great area of the association cortex, which forms 95% of the neocortex (cf. Figs. 1-1, 3-1). In Fig. 2-6 there are shown on each side the discs of termination of the thalamo-cortical afferents (Spec. aff) in lamina IV, which are out of register with the central module structured around the termination of the cortico-cortical (association) afferent.

The sequential relays from the primary to the secondary to the tertiary, and so on, areas have been studied by Jones and Powell (1970b) by a technique of successive degeneration. Figure 3-2 is constructed from these findings to show the connectivities from the primary areas for somaesthesis and vision. A similar diagram could be made for hearing. The principal pathways from primary to secondary to tertiary are shown by thickened arrows, the numbers corresponding to the Brodmann areas (Fig. 3-1) with some additions defined in the legend. The connectivities shown in Fig. 3-2 are of particular interest to our theme of conscious perception because they represent the way to the elaborated spatio-temporal patterns of neuronal activity that are the necessary substrate of conscious perceptions. They are also of interest because they represent the first stages of the modular relay by association fibres that was discussed in Lecture 2 (cf. Fig. 2-5) in relation to the connectivities within the association cortices.

3.3 Physiological Investigations

Transmission from the columns of areas 3, 1, 2 out to the somaesthetic association areas, 5 and 7 (Fig. 3-2), has been shown by Mountcastle et al. (1975) to be in the modular mode. There has been a most sophisticated investigation of the responses of neurones in monkeys which were subjected to a variety of testing situations. In that way

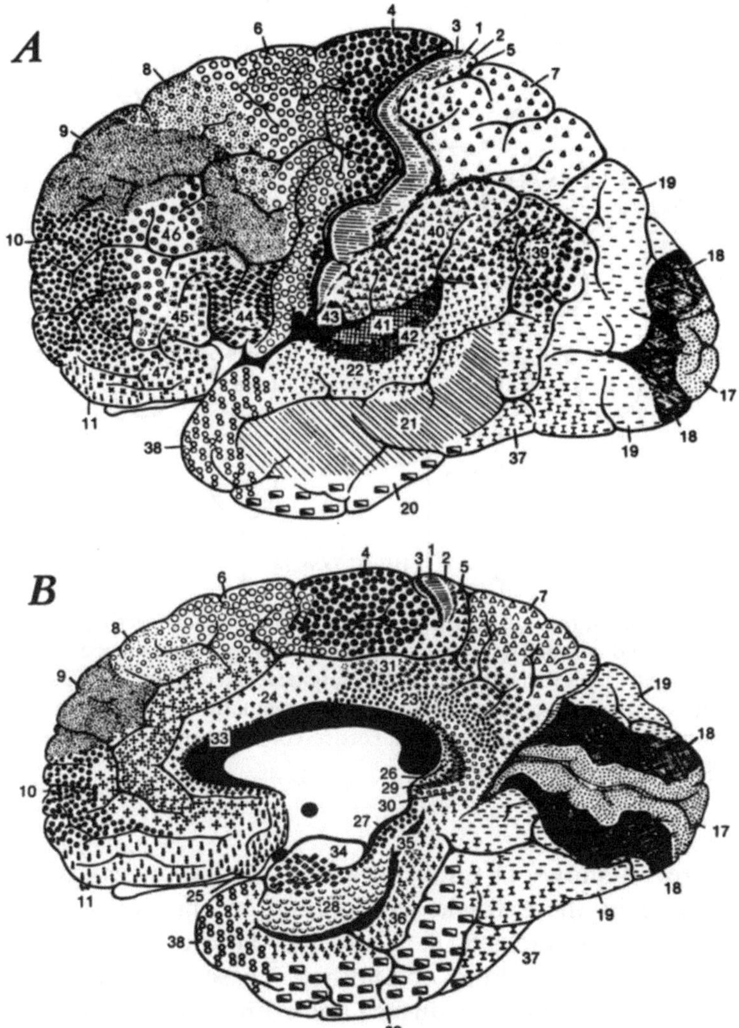

Fig. 3-1 A, B. Brodmann's cytoarchitectural map of the human brain. The various areas are labelled with different symbols and their numbers indicated by figures. **A** Lateral view of left hemisphere. **B** Medial view of right hemisphere. Brodmann, K.: Vergleichende Lokalisationslehre der Großhirnrinde. Leipzig: Barth 1909

neurones were found to belong to as many as seven categories in areas 5 and 7. The neurones of any one set were arranged in vertically oriented columns or modules, but the structural relationships of these modules were not clearly defined.

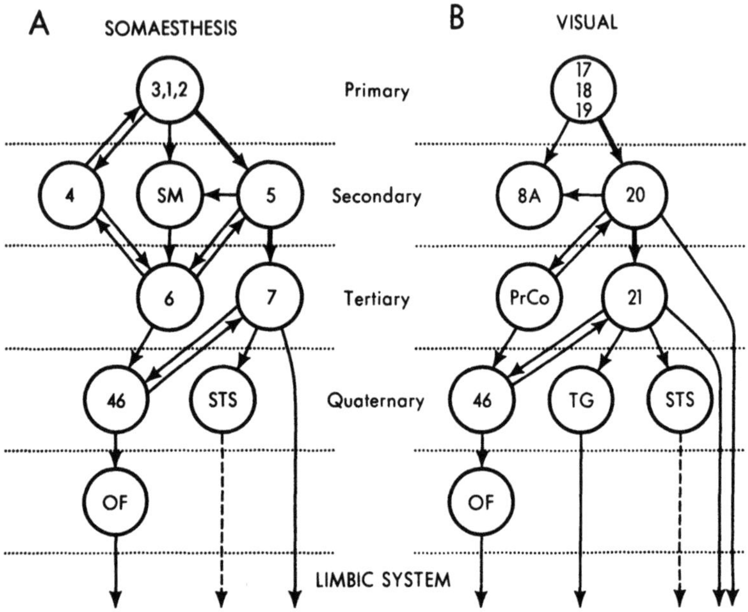

Fig. 3-2A, B. Diagrammatic representation of cascade of connectivities for the somaesthetic (**A**) and visual (**B**) systems in the cerebrum. The numbers refer to the Brodmann areas; the other areas are: *SM*, supplementary motor; *STS*, superior temporal sulcus; *PrCo*, precentral agranular; *TG*, temporal pole; *OF*, orbital surface frontal lobe. All the *arrows* represent cortico-cortical connectivities, the *thickened arrows* show the principal pathways

Figure 3-3 shows 11 tracks in area 7. The four sets of neurones encountered in these tracks are plotted in the code given below the tracks. With tracks 5, 8, 9, 10 and 11 the neurones belonged to a single set of one kind or another. These were all short tracks more or less normal to the surface and so presumably were confined to one module. On the other hand tracks 1, 2, 4, 6 and 7 were more oblique and consequently two or more different sets of neurones were encountered. The odd neurones not belonging to the sets are indicated by horizontal bars. Mountcastle et al. (1975) do not derive an estimate for module size from Fig. 3-3 and a more complicated figure for area 5, but it seems that a cross-section of about 0.5 mm would be a reasonable estimate. Thus in these initial relays from the somatosensory cortex there is a modular structure not too different from that observed by Goldman and Nauta (1977) in association cortex connectivities. Investigations by radiotracers or horseradish

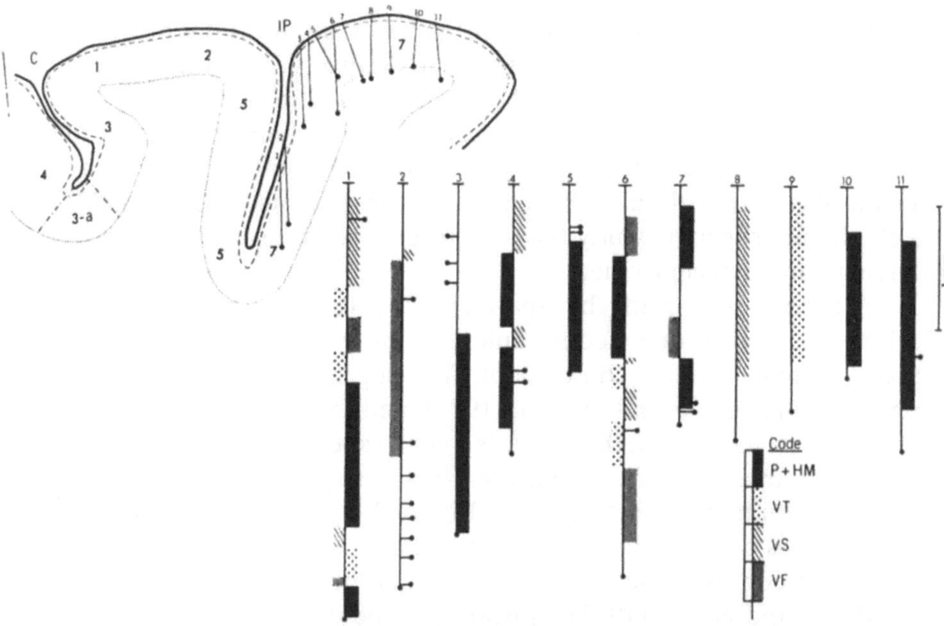

Fig. 3-3. A sample of reconstructions of microelectrode penetrations made into area 7 of waking, behaving monkeys in the present experiments. Code to *lower right* indicates classes of neurones encountered: *P+HM*, projection and hand manipulation neurones; *VT*, visual tracking neurones; *VS*, visual space neurones; and *VF*, visual fixation neurones, to left passively, to right actively. The *horizontal bars* indicate loci at which single neurones were isolated which were out of place with regard to the en bloc pattern of columnar organization; they are all neurones of the active and special classes and suggest that a dynamic mode of organization is at times superimposed on that of columnar organization. Mountcastle, V.B., Lynch, J.C., Georgopoulos, A., Sakata, H., Acuna, C.: Posterior parietal association cortex of the monkey: command functions for operation within extrapersonal space. J. Neurophysiol. *38*, 871–908 (1975)

peroxidase would be valuable in providing more reliability in the modular structure.

In the visual system of the cat Hubel and Wiesel (1965) made a detailed study of Brodmann areas 18 and 19 to which the primary visual area 17 projects. Both these visual areas were found to have a columnar arrangement of neurones with similar properties, particularly in respect of their receptive fields. There appeared to be an intense interaction between the neurones of a column and most were activated from both eyes. There was no estimate of the cross-sectional area of a module, but from the published results with oblique tracking it would appear to be of the order of 0.5 mm or less.

In the monkey, Zeki (1974, 1976) has found that in the projection from areas 17 and 18 there were two well defined areas in the superior temporal sulcus (STS in Fig. 3-2B). There was a rather crude columnar representation as studied by recording from single cells. In one area the cells were especially sensitive to movements of objects and in the other to colour with a remarkable specificity (Zeki, 1977). Zeki did not specially study the topographic relations of the neurones having similar receptivities.

A topographic study by Spatz (1977) gives the first clue on the initial stage of modular communication by association fibres of the visual system. Using techniques of fibre degeneration, autoradiography (cf. Goldman and Nauta, 1977) and horseradish peroxidase transport he has shown that there is a sharp specificity of communication from area 17 to the visual area in the superior temporal sulcus that was studied by Zeki (1974). Figure 3-4 shows the topography of the zones of communication, there being a strict spatial relationship. The connectivities are coarser in grain than those found by Goldman and Nauta (1977), appearing to be 0.5 to 1.0 mm diameter

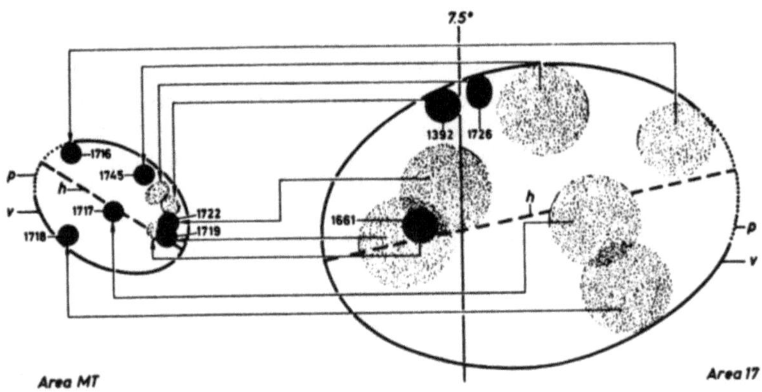

Fig. 3-4. Schematic drawing of unfolded area 17 and area MT showing the topography of the projection of area 17 upon area MT. *Black areas* indicate operation sites (lesions and [^{3}H]proline injections in area 17, HRP injections in area MT) which are connected by thin lines with the sites of the respective reaction product (*stippled areas*). *Arrows* indicate direction of projection, not of transport of tracers. The *straight line* through area 17 separates schematically the lateral portion of area 17 (central vision out to 7.5°) from its medial portion. *h*, horizontal meridian; *p*, extreme periphery; *v*, vertical 0-meridian. Spatz, W.B.: Topographically organized reciprocal connections between areas 17 and MT (Visual area of superior temporal sulcus) in the Marmoset Callithrix Jacchus. Exp. Brain Res. *27*, 559–572 (1977)

for a module. However it has to be realized that with the techniques used several adjacent modules may be injected with a consequent coarsening of the observed grain in Fig. 3-4. At least we can say that in the visual system the association communication out from the primary sensory area is constituted by a specificity of connectivities such as has been demonstrated for the association cortex (cf. Figs. 2-4, 2-5).

3.4 Psychological Comments

Before embarking on the complex psychological problems of visual illusions, some general psychological considerations may be helpful.

It is relevant to quote from Popper (Popper and Eccles, 1977, Sect. 13):

> 'Our visual perception is more like a process of painting a picture, selectively (where "making comes before matching" as Ernst Gombrich says) than one of taking random photographs': and later in Sect. 15: 'we should look upon World 2 as active – as productive and critical (making and matching)'.

These ideas on visual perception are well illustrated by the eye movements observed when gazing at a picture (Yarbus, 1967). With a portrait, for example, features of special interest such as the eyes, lips and nose are returned to again and again in the saccadic eye movements, yet the picture is perceived as a steady object (Fig. 3-5).

Sutherland (1968) outlined a psychological theory of visual pattern recognition which in essentials resembled Gombrich's (1960) ideas of making and matching. His discussion provides a good illustration of the formidable difficulties in the physiological and psychological problems that arise even in the recognition of a simple pattern.

3.5 Visual Illusions

3.5.1 General Considerations

It is generally recognized that the study of visual illusions can be of importance in the attempt to understand how visual perceptions

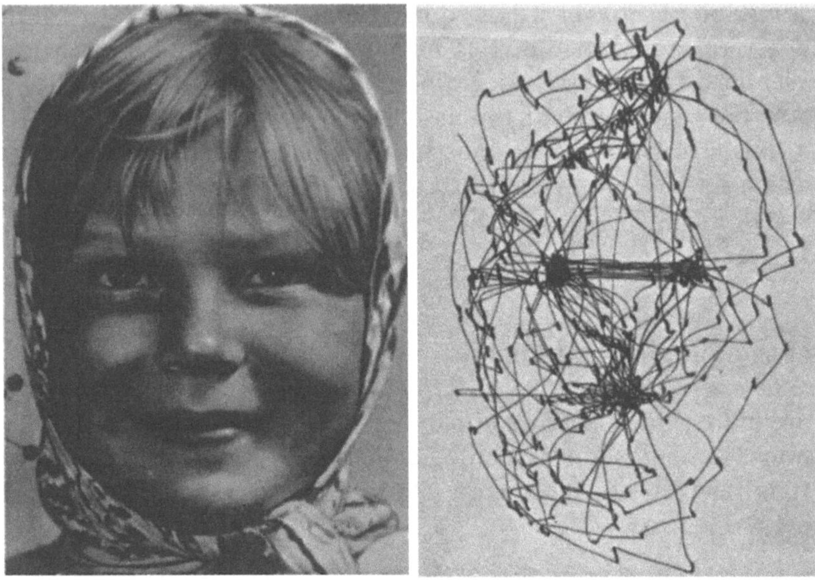

Fig. 3-5 A, B. Regularities of eye movement appear in a recording of a subject viewing a photograph of a child. **A** What the subject saw; **B** His eye movements as recorded. The eyes seem to visit the features of the head cyclically, following fairly regular pathways, rather than criss-crossing the picture at random. Yarbus, A.L.: Eye movements and vision. New York, Plenum Press 1967

are developed from the retinal images. As has been well stated by Gregory (1973):

> Many illusions of perception are more than mere errors; they may be experiences in their own right. They can illuminate reality. It is this power of illusion to illuminate which the artist somehow commands. To the scientist illusions present a curious challenge; for though phenomena, they are not phenomena of his physical world – they are phenomena just because they are deviations from what he takes to be reality.
>
> In certain situations the eye and brain are confounded, to generate illusions. Illusions may be mere errors; or they may be frightening or fascinating unearthly experiences, which may stimulate the imagination to seduce us from reality. Some illusions are irritating or dangerous; but others are at least as interesting and as much fun as reality.

Illusions probably occur in all the senses, but have been most effectively studied in the visual system on which we will now concentrate.

In the first stage leading to visual perception an image of the external world is formed on the retina as a two dimensional (2D)

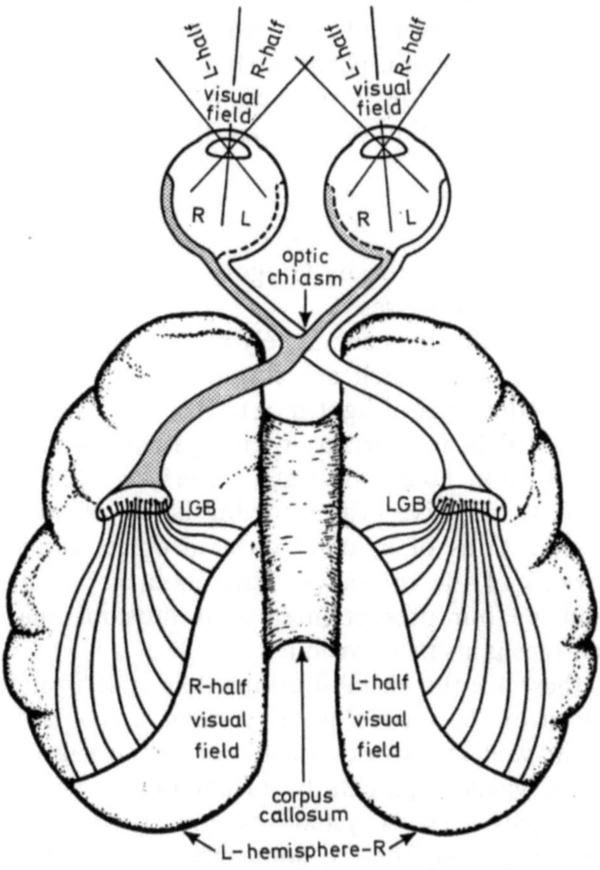

Fig. 3-6. Diagram of visual pathways showing the left-half and right-half visual fields with the retinal images and the partial crossing in the optic chiasma so that the right-half of the visual field of each eye goes to left visual cortex, after relay in the lateral geniculate body (*LGB*), and correspondingly for left visual field to right visual cortex. Popper, K.R., Eccles, J.C.: The self and its brain. Berlin, Heidelberg, New York: Springer 1977

picture (Fig. 3-6). Because of the separation of the optic axes of the two eyes by 60 to 70 mm, there is a slight disparity in the two retinal images, which causes the phenomenon known as binocular parallax. Near objects are shifted relative to far, as can be easily seen by observing first with one eye then with the other. This disparity is important in building perceptions of depth, a 3D perception arising from the two slightly different 2D images on the retinas (cf. Sect. 3.6). But perceptions of depth can be experienced by looking at a painting, which is a 2D object observed with no retinal disparity, and yet we have an illusion of depth, particularly if the painting is observed with one eye only so that the 3D perception can arise without conflict by the binocular information of zero parallax.

3.5.2 Classification

We can accept the subdivision by Gregory (1968) of visual illusions into three quite distinct classes.

Firstly, there are the illusions completely accounted for by the simple physical principles, and which are commonplace experiences. The commonest are reflections of light in a mirror. Other examples arise from refraction of light in a prism or lens system, the simplest being the bent-stick illusion at an air–water interface. Related illusions are mirages due to selective heating giving layers of air of different densities, or refractions of light by water droplets resulting in the beautiful illusions of the rainbows. This class of visual illusion is completely outside our present interest in illusions that relate to the perceptual mechanisms of vision.

Secondly, there are the visual illusions that arise from distortions in the operation of the physiological systems concerned in vision. For example the well known experiences of after-images result from the bleaching of visual pigments by bright lights. Then there is in the retina a complex nervous system that greatly distorts the signals sent to the brain in response to some retinal illumination. Some retinal ganglion cells signal to the brain by impulse discharges when the associated retina is illuminated, others signal darkness. Most ganglion cells signal at the beginning and end of a retinal illumination. So the retinal nervous system is signalling to the brain differences in illumination in time and space. The many illusions recognized as contrast phenomena arise in this way, for example the Mach bands at a light–dark interface (Ratliff, 1972).

These various distortions result in the input of misleading information into the visual areas of the cerebral cortex. However, there also can be in the visual cortex further modifications of the sensory input that result from complex interactions in the neural machinery. Cells are responsive particularly to lines and edges of the light stimuli applied to the retina with specificities for orientation and for direction of movements across the visual field. Jung (1973) has given an authoritative account of the physiological processes concerned in a variety of these physiological illusions, for example, the Hermann grid with contrast illusion, Ehrenstein's brightness illusion and the Craik-Cornsweet illusion of border-induced brightness and darkness. There is good correlation with the observed neuronal responses, and explanations are given in terms particularly of inhibitory interactions of cortical neurones in accord with the diagrams of Figs. 2-2 and 2-6. Blakemore (1973) also gives a well illustrated account of visual illusions that can be attributed to such physiological mechanisms in the visual cortex. For example the after-effects following concentrated gaze on gratings of various kinds and orientations result in fascinating visual experiences. Relatively simple physiological explanations are offered in terms of fatigue of detector neurones and of the inhibitory interaction between different sets of detector neurones.

Thirdly, there are perceptual or cognitive illusions in which there is a misinterpretation or an ambiguity of interpretation with respect to the operations of the visual machinery of the cerebral cortex. It is important to recognize that these illusions arise in the perception of the visual inputs. They are strictly a phenomenon of conscious experience. We may ask how related are such illusory experiences to the operations of the neural machinery of the cerebral cortex. In fact the chief interest of these cognitive illusions lies in the light they may shed on the brain–mind problem. Some of these illusions are clearly cognitive, but with others there may be both a physiological and a cognitive basis. It is important to realize that cognitive illusions are experienced by a subject that has had a long discipline in learning to interpret visual inputs. As far as is known, they are not experienced by a very young child or by an older person to whom sight has just been given by an appropriate operation, for example the removal of a congenital cataract. Gregory (1972) reports that claims have been made that pigeons and even fish can be trained to give responses indicating simple cognitive illusions arising from errors in judgement of lengths of lines. However he is somewhat sceptical!

3.5.3 Cognitive Illusions

3.5.3.1 Errors of Judgement

All of these errors arise from misinterpretation of one set of operations of the cortical machinery because of some conflict with other experiences or with remembrances. There is evidence (Gregory, 1972) that physiological explanations are insufficient; hence the illusions presumably arise in malfunctioning of the brain–mind interaction.

(1) Size constancy illusion (Gregory, 1972). Familiar objects are seen to be much the same size whether near or far, yet in the whole physical and physiological mechanism for seeing, the size is inversely proportional to the distance. A striking example is to look at your two hands, one held near the face, the other at full arms length. They appear to be the same size, yet, when the near is moved across to overlap the far, it grows in size, i.e. the size constancy illusion is negated. The maintenance of apparent size constancy despite increasing distance is related to an effect known as Emmert's law. It has been shown by Gregory (1968) to be due to a central scaling mechanism that allows for changes in apparent distance during the process of perceptual interpretation of the cortical responses evoked by the retinal image.

(2) Müller-Lyer Illusion. In Fig. 3-7A the shafts of the two figures actually are the same length, yet that with outgoing arrowheads appears to be much longer than that with ingoing heads. In Fig. 3-7B the illusion surprisingly remains for the arrowheads without the shafts. As a result of ingenious investigations Gregory (1972) has eliminated several suggested explanations. He proposes an explanation based on perspective scaling and the size constancy illusion. Perspective indicates that the outgoing arrow figure is an internal corner of a room and the ingoing arrow an external corner, which would be expected to be nearer. The outgoing arrow corner being more distant would be expected to be shorter, hence the illusion of a perceived length longer than for the external corner.

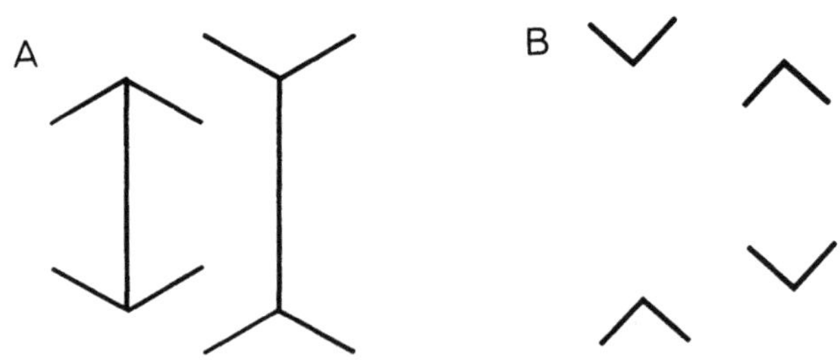

Fig. 3-7 A, B. The Müller-Lyer illusion. **A** The shaft with inward going *arrowheads* appears shorter than that with outward going *arrowheads*. Gregory, R.L.: Perceptual illusions and brain models. Proc. R. Soc. Lond. B *171*, 279–296 (1968). **B** The illusion appears even in the absence of the shaft. Gregory, R.L.: Eye and brain, pp. 251. London: World University Library 1972

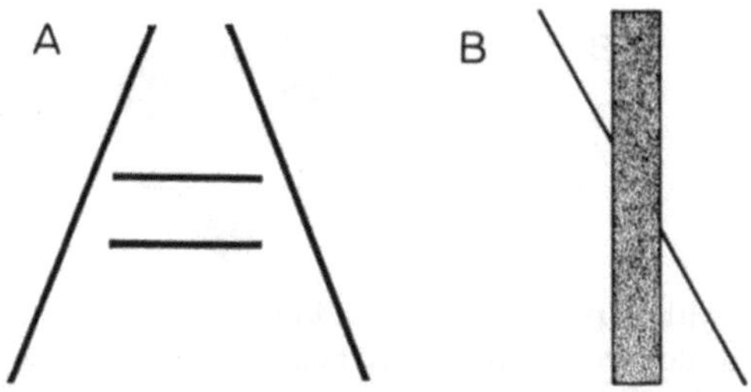

Fig. 3-8. A The Ponzo illusion. The upper of the parallel lines is expanded with respect to the lower. **B** The Poggendorff illusion. The straight line crossing the shaded rectangle appears displaced. Gregory, R.L.: Perceptual illusions and brain models. Proc. R. Soc. Lond. B *171*, 279–296 (1968)

(3) The Ponzo illusion (Fig. 3-8 A) can be simply explained by seeing the two sloping lines as a receding railway track, the upper line being judged longer because of its greater distance.

(4) The Poggendorf illusion (Fig. 3-8 B) is the earliest geometrical illusion to be reported. The sloping straight line appears to be displaced as it crosses the rectangle.

(5) There are many illusions having this character of distortion of lines crossing at angles on some background of parallel lines. They are distinguished by the names of their discoverers – Hering, Wundt, Zollner, Orbison – and all are illustrated by Gregory (1972).

(6) Another illusion arising from errors in judgement is the well known moon illusion. The moon appears to increase in size as it nears the horizon. Another variety occurs in the ingenious rooms designed by Ames. They have walls, floors and windows departing far from the rectangular design we are accustomed to, yet when seen from one viewing site they appear to be rectangular and subjects standing in various situations are as a consequence judged to be greatly changed in size – even to be doubled or halved in height.

3.5.3.2 Completion Illusions

It is well known by artists that when given a minimum outline drawing, the viewer will complete interrupted lines and fill out his perception to gain a remarkably rich experience. It adds to the delight of the viewer thus to participate with the artist in creating a full perceptual experience coherent with the artist's outline. This performance is a measure of the trained mind of the viewer. This completion ability can be recognized in the very simple designs of Fig. 3-9 (Gre-

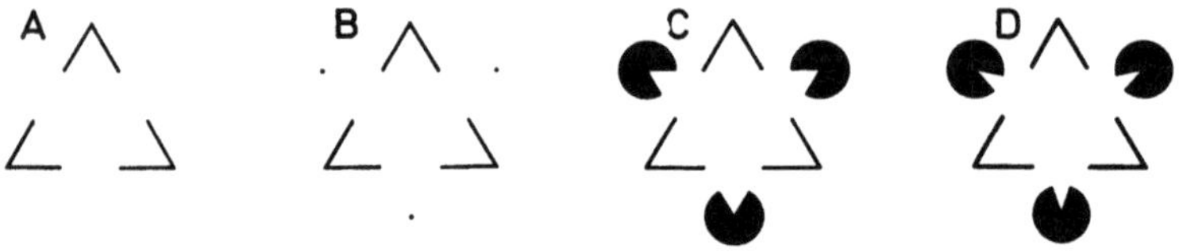

Fig. 3-9. A Triangle with gaps. **B** With the addition of dots a faint illusory triangle is visible. **C** With the dots replaced by a strong indication of the apices, the illusory triangle is enhanced. **D** With the indication of narrower apices the illusory triangle develops curved sides. Gregory, R.L.: The confounded eye. In: Illusion in nature and art. Gregory, R.L., Gombrich, E.H. (eds.), pp. 49–95. London: Duckworth 1973

gory, 1973). The triangle with gaps (A) receives the superposition of a faint illusory triangle when three dots are added (B). This illusion is greatly strengthened by the definition of its angles (C). By narrowing the apices (D) an illusion is created of a concave sided triangle that does not fit the gaps in the sides of the basic triangle. These figures illustrate illusions that are created on the basis of minimum input signals.

3.5.3.3 Illusions Characterized by Ambiguity

The most famous ambiguous illusion is the Necker Cube (Fig. 3-10A). It is seen globally in either of two configurations. The face with the small circle in the centre may be either in front or behind. The switch from one configuration to the other seems to be instantaneous and not dependent on any change in the viewing angle. An even more striking global switching occurs for the illusion which

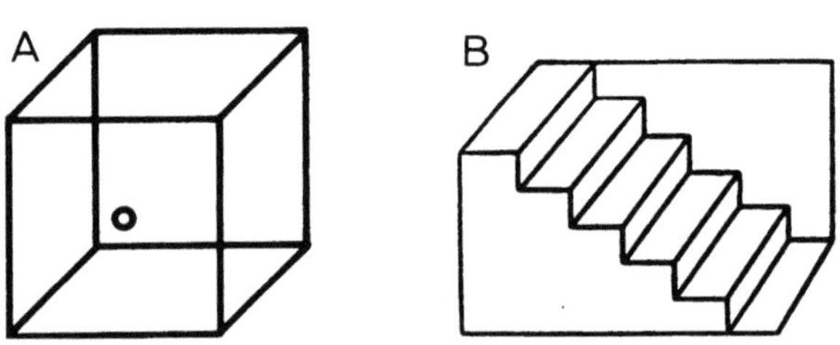

Fig. 3-10. A The Necker cube, the most famous of the many depth ambiguous figures. The circle in the centre of one face aids in the identification of this face when it is either in front or behind. Gregory, R.L.: Eye and brain, pp. 251. London: World University Library 1972. **B** The staircase – cornice ambiguous figure. There is similarly a reversal of near and far faces

may be either an ascending stair or an over-hanging cornice (Fig. 3-10B). With most observers the staircase is the dominant illusion, presumably because staircases are much more commonly observed than cornices. Again no transition stage is ever experienced.

There are several other ambiguous figures displaying two alternative interpretations that also exhibit global switching: young girl or old woman; Indian or Eskimo; rabbit or duck (cf. Attnaeve, 1974; Gombrich, 1973). The most remarkable are in the drawings of Escher that have been so penetratingly discussed by Marianne Teuber (1974), where with great ingenuity he has filled the entire picture. Heaven and Hell, Day and Night and Circle Limit 1 give fascinating examples in which there is a global reversal of figure and ground. An interesting example of an ambiguous figure is the Rubins vase (Fig. 3-11) (Gombrich, 1973). If attention is concentrated on the flowers above, it is a vase, but global reversal to the double profile occurs if attention is switched to the two ears at the side. Figure 3-12 (Altneave, 1974) shows a remarkable figure–ground reversal because in the reversal there has been a synchronous change in the directional orientation by 90° corresponding to the shapes of the units of the pattern.

3.5.3.4 Attempted Explanations of Cognitive Illusions

3.5.3.4.1 Introduction

Cognitive illusions raise great difficulties for any attempt at an explanation which is restricted to the operations of the neural machinery with the conscious mind in some identist or parallelist relationship. Gregory (1972, 1973) attempts to formulate some way to approach an explanation of cognitive illusions, but it suffers from the impossible handicap that he restricts his considerations to brain states. For example (Gregory, 1973, p. 91): 'If brains can describe states of affairs or possible states of affairs, the descriptions may be logically impossible, though the brain mechanisms are not.'

If for brains we substitute 'minds', this statement is close to the general hypothesis of cognitive illusions that I shall propose. And again (Gregory, 1974, p. 52): 'The Necker cube is an example of a picture in which the depth ambiguity is so great that the brain never settles for a single answer'.

In reference to the features selected by the physiological mechanism, for example lines and angles, Gregory (1973) very pointedly asks: 'How are these selected features, as neurally represented, combined to give perception of objects?'

He rightly rejects the absurd hypothesis that there are single cells specifically tuned to any object that we can recognize. But he has no alternative, merely stating: 'it is generally assumed that many signal features are *somehow* pieced together, to build up perceptions of objects from available data' (my italics) (Gregory, 1973). Also Gregory (1974, p. 57) states that: 'we can think of perception as being essentially the selection of the most appropriate stored hypothesis according to current sensory data'.

I am in complete agreement if the hypothesis is in the mind and if the selection is guided by the search for meaning.

In a beautifully illustrated and thoughtful paper on 'Illusion and Art', Gombrich (1973, p. 240) stresses the search for meaning in perception:

> Those who ask about our 'beliefs' in front of paintings are certainly asking the wrong question. Illusions are not false beliefs, though false beliefs may be caused by illusions. What may make a painting like a distant view through a window is not the fact that the two can be as indistinguishable as is a facsimile from the original; it is the similarity between the mental activities both can arouse, the search for meaning, the testing for consistency, expressed in the movements of the eye and, more important, in the movements of the mind.

The reference to the distant view through a window recalls Margritte's picture *La condition humaine*, 1933, where there is shown a painting partly overlapping the view through a window and being an excellent facsimile.

In a very well illustrated and argued contribution, *The perception of disorientated figures*, Rock (1974, p. 78) concludes that

> form perception in general is based to a much greater extent on cognitive processes than any current theory maintains ... Although the work I have described does not deny the possible importance of contour detection as a basis of form perception, it does suggest that such an explanation is far from sufficient, and that the perception of form depends on certain mental processes such as description and correction. These processes in turn are necessary to account for the further step of recognition of a figure.

Attneave (1974, p. 98) comes to a related conclusion in his attempt to explain the perception of ambiguous figures and their reversal, in that he proposed: 'the existence of some kind of working model of three dimensional space within the nervous system that solves problems of this type by analogue operations'; though he admits that there is no neural organization that could subserve such a model. However I will suggest below that this model is in the mind, not in the brain. These suggestions of Rock and Attneave lead on to the hypothesis of perception that I shall now proceed to develop on the basis of a dualist–interactionist formulation of the brain–mind problem.

It is proposed that in the awake state the mind is continually scanning or probing the spatio-temporal patterns of the modular activities of the cerebral cortex (cf. Figs. 1-7, 2-10) in accord with its interests and attention. This activity itself is not consciously perceived, being analogous to the 'unconscious inference' postulated by Helmholtz that is referred to in the above quotation from Jung (1978). The deliverances of this reading out reach consciousness as perceptions. Thus the piecing together of the immensely complex patterns of the neural machinery would be the function of the mind acting unconsciously in accord with its remembrances and its present interest to achieve a synthesis from moment to moment that is transmuted into the sequences of conscious perceptions. This subconscious operation is always searching for meaning. We are not much interested in multitudinous arrays of visual data that have no significance or meaning. Certainly when one considers the evolutionary origin of our visual perception, it was meaning that was paramount for survival. Even if the meaning was subject to error, it was preferable to no meaning. And so we have here in simple outline an hypothesis of perception that has great explanatory power with respect to illusions. In fact we can regard illusions as providing tests for the validity of the hypothesis.

3.5.3.4.2 Errors of Judgement

The size constancy illusion is readily explained on the basis of remembered knowledge – e.g. of hands being the same size, or of people being the same height regardless of their apparent diminution with distance. The Müller-Lyer and the Ponzo illusions (Figs. 3-7A, B; 3-8A) presumably relate to an association with perspective imagery as proposed by Gregory, and this would also obtain for the Ames illusion of size. No such simple explanation can be offered for the Poggendorf (Fig. 3-8B), Hering, Wundt, and Zollner illusions, which may have in part a physiological explanation as surmised by Gregory

(1972). Possibly the moon illusion arises from the well known cognitive tendency to amplify images close to the horizon, as is also illustrated in our exaggerated perception of distant mountains.

3.5.3.4.3 Completion Illusions (Fig. 3-9)

For effective seeing under favourable conditions one is all the time trying to discover and identify objects that are revealed by inadequate visual data. This is a skilled cognitive performance of great significance in such operations as hunting and warfare. The moving picture *Blow Up* was essentially on this theme. As Popper states (Popper and Eccles, 1977, p. 430):

> There is an incoming challenge from the sensed world which then puts the brain, or ourselves, to work on it, to try to interpret it ... to match.

We are all the time attempting to make more out of a visual experience than is actually given, and so to give it meaning. We work by postulates or guesses that are consciously tested as far as possible by critical judgement and by the continuing input of new data. And in hunting and warfare and exploration these guesses are checked by cross-reference to other observers. All of this is a complex cognitive performance based on the conscious perceptions being read out from and synthesized from the spatio-temporal patterns of the modular activities in regions of the neocortex. Moreover this interaction of mind with brain is indicated in Fig. 1-7 as a two-way performance mind$\rightleftharpoons$brain that is continuously in operation in the quest for meaning. On the basis of these general considerations it is not difficult to give a satisfactory explanation of completion illusions.

3.5.3.4.4 Illusions Characterized by Ambiguity (Figs. 3-10, 3-11, 3-12)

The dualist-interactionist hypothesis of perception is particularly effective in relation to ambiguous figures. For example with the Necker cube or the staircase–cornice illusions, the visual input results in some spatio-temporal pattern of modular activity (cf. Fig. 2-10). In the process of reading out by the subconscious to the conscious mind the attempt is made to give a holistic meaning to the geometrical

Fig. 3-11. The Rubins Vase. According as attention is focussed on the flowers above or the ears at the side there is a reversal in identification. Gombrich, E.H.: Illusion and art. In: Illusion in nature and art. Gregory, R.L., Gombrich, E.H. (eds.), pp. 193–243. London: Duckworth 1973

Fig. 3-12. Reversal and rotation occur simultaneously in this ingenious design. When the stylized maple-leaf pattern alternates between black and white, it also rotates 90 degrees. Attneave, F.: Multistability in perception. In: Image, object and illusion. Held, R. (ed.), pp. 90–99. San Francisco: Freeman 1974

construction that is represented in coded form in the spatio-temporal patterns of modular activity. However there are two quite distinctive meanings. The strategy of the alternation of interpretation is subconscious. So we are given, apparently at random, first one interpretation then the other. The two possible meanings are given holistically, never fragmented. Moreover the switch appears to be instantaneous. A somewhat related stereopsis illusion takes about 40 ms for development (Julesz, 1971) as will be described below. So there could be this order of delay in switching between the two interpretations of the ambiguous figures. Such a switching time would be too short for introspective detection. The same explanation of mental readout of meaning would account satisfactorily for the other ambiguous figures (Figs. 3-11 and 3-12).

Identity theorists and parallelists are confronted by the grave problems of explaining how the same brain events can appear as two identities in perception. The quotations given above from Gregory, Rock and Attneave indicate the dilemma confronting those attempting to give an explanation on some variety of the identity theory. For dualist–interactionism the duality of experience is attributed to the search for meaning and the finding that there are two meanings compatible with the neural events.

3.6 Depth Perception

The perception of depth in the perceived image can be regarded as an illusion because there is only a 2D image on the retina, yet it is interpreted as being derived from a 3D object. When we enquire how this illusion is created, we can find many answers when considering the rich visual experience from the perceived world. We will concentrate firstly on stereopsis which is given by stereoscopic vision because it is of great significance and has been intensively investigated.

3.6.1 Stereopsis

As already stated, depth perception is principally dependent on disparity in the retinal images because of the separation of the eyes by 50 to 70 mm. By using random dot stereograms Julesz (1971, 1974)

Fig. 3-13. Random-dot stereogram that can be interpreted to have a raised central square when viewed stereoscopically either by a stereoscope or by 'crossing the eyes'. Julesz, B.: Foundations of cyclopean perception, pp. 406. Chicago: University of Chicago Press 1971

has made a most thorough study of the visual input required for stereopsis, that is for perception of depth in abstract patterns. These random dot stereograms were generated by a computer, and a central area of one frame was displaced laterally by several dots relative to the other frame. When viewed stereoscopically or by 'crossing the eyes', this area of the pattern of Fig. 3-13 is perceived as a square floating in space above the background. This shows that spatial discrepancy of the two retinal images is sufficient for this stereoscopic effect. The distance of the stereoscopic effect is dependent on the lateral separation in the stereogram. By reversing the stereopsis the central area is observed to be depressed below the surround.

In the simplest stereograms under the best conditions stereopsis is secured within 50 ms. However with the most subtle and complex stereograms it may take many seconds or even minutes to secure stereopsis, which can then be rewarded by magically beautiful scenes in depth. It is interesting that, once stereopsis has been achieved with a difficult stereogram, it can be much more readily established on later presentations, which can be regarded as a learned effect. Also verbal descriptions of the image to be perceived in depth aids in its recognition.

A subtle technique was developed by Julesz in 1965 for masking the after-image of the initial presentation of a flashed stereogram by a second flashed stereogram similar to the first except for its ambiguous properties with the same disparity in both directions. It was found that stereopsis was secured when the presentation interval was as brief as 40 ms, which can be regarded as the minimum perception time for stereopsis. Later (Julesz, 1971) the time was found to be as brief as 10 ms under the most favourable conditions. This time can be regarded as the minimal operating time for the neural machinery to provide the coded information necessary for the perception of stereopsis. A stereopsis time of less than 50 ms and even down to 30 ms is indicated by random dot stereo-movies that can generate a perceived movement even at 33 frames/s (Julesz, 1971).

By using the exhibition of illusory figures in random dot array it has been demonstrated that some of the illusions disappeared or were greatly diminished when the constituent parts were seen at different levels. This is observed for the Ponzo, Ebbinghams, Poggendorf and Zollner illusions. However the Müller-Lyer illusion (Fig. 3-7 A, B) can still be fully observed in the random dot stereopsis, which indicates that it is generated by a higher central mechanism than the other illusions of judgement.

Investigations of random dot stereograms are important because they demonstrate how the illusion of depth perception can be experienced in the absence of any other clues. The physiological counterpart has been displayed in elegant studies by Bishop (1970); by Barlow et al. (1967), and by Hubel and Wiesel (1970) on single neurones in the visual cortex that signal binocular disparity. Such neurones respond differentially to overlapping visual inputs from the two eyes. Barlow et al. (1967) have demonstrated various gradations of this disparity response, which can be regarded as a specific signalling of depth perception of varying degree (cf. Poggio and Fisher, 1977). A sophisticated theory of the events responsible for stereovision has been developed by Marr and Poggio (1979) on the basis of disparity neurones. It involves an incessant and critical matching process together with memory storage. It is very satisfactory to have this most challenging development of stereopsis theory, but as yet it is not related to precise neuronal mechanisms.

3.6.2 Other Factors Concerned in Depth Perception

As soon as he became interested in painting with some approach to verisimilitude, the pictorial artist was confronted with the problem of creating a flat picture that gave the viewer an illusion of depth. In his Mellon Lectures, 'Art and Illusion', Gombrich (1960) presents a most learned and subtle account excellently illustrated. A list of all the devices adopted by artists is compiled by Blakemore (1973) with appropriate illustrations. For my present purpose a list is sufficient because as Blakemore states: 'it is impossible to comprehend how the monocular perception of distance is achieved'. By that he means that there is no physiological explanation of the illusion of depth created by the following devices used by artists:

a) Position in the field, a primitive device, the further an object is away the higher in the picture.
b) Perspective, which was the special achievement of the Renaissance artists.
c) Texture gradient, that becomes finer with distance.
d) Shadow gives indication of relative position of objects in the field.
e) Overlay gives unequivocal evidence of relative positions of objects.
f) The plein-air effect, which results in haziness and bluer colour of distant objects.

It must have been a discouragement to pictorial artists to find that, despite their best efforts, the picture still failed to take on the full depth of the observed scene. The reason is of course that there is no binocular disparity, hence there is the powerful counter-illusion that the picture is flat. Correction is easy, yet very rarely applied. We have to view all pictures of scenes monocularly. It is sufficient to cover one eye in order to experience the magical transformation of a picture in 2D into a scene in depth (3D). As recounted by Gombrich (1973) this can simply be done by monocularly viewing a picture through a wide tube. In that way by elimination of parallax one can realize the effectiveness of the above seven strategies for giving the illusion of depth to a flat picture.

3.6.3 Impossible Figures

These are really out of place in an account of illusions. Yet they are related because they illustrate the challenge to make 3D sense of 2D drawings that *appear* to represent a 3D object. For example the drawing of Fig. 3-14A cannot be seen as an object in three

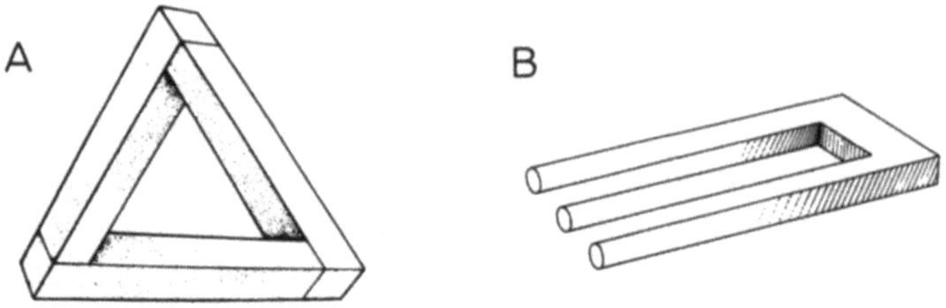

Fig. 3-14 A, B. Impossible figures. **A** This drawing cannot be seen as a three dimensional structure. Gregory, R.L.: Perceptual illusions and brain models. Proc. Roy. Soc. *B 171*, 279–296 (1968). **B** This drawing gives the illusion of depth of two structures which are mutually irreconcilable. Gregory, R.L.: Eye and brain, pp. 251. London: World University Library 1972

dimensions, and so it presents a greatly disturbing problem to our interpreting mind. If any one of the corners is covered up there is no problem because one of the arms immediately moves out of the plane defined by the other two. Another disturbing drawing in Fig. 3-14B presents no trouble if either end is covered up. It becomes either a double pronged structure or three cylinders. More complex impossible figures were created by Escher – Belvedere, Concave and Convex, and Relativity are all illustrated by Marianne Teuber (1974). She points out that Escher denied that he had developed these extraordinary pictures from Piranesi who had constructed impossible pictures in the eighteenth century.

3.6.4 Attempted Explanations of the Illusions of Depth Perception

Though there have been remarkable investigations on depth perception in stereopsis and on the neurones signalling disparity, no coherent explanation of the illusion itself has been forthcoming. This is to be expected because in materialist terms it has not been possible to give an account of how the immense diversity of neuronal responses coding some visual input is synthesized to give a perceived picture. This inadequacy of the identity theory in all its formulations has already been demonstrated in Lecture 2.

However, on the dualist-interactionist hypothesis it is possible to give a general account of how the visual information encoded in the spatio-temporal patterns of modular responses comes to be transmuted into a perception. It has to be recognized that, in the world we live in, depth perception is a learned response requiring years of experience for its full development (cf. Chap. 7, Eccles, 1979a). Learned experience is also of great significance in deriving the perception of depth in pictures, as can be well illustrated by the deficiencies in this respect of primitive peoples (cf. Deregowski, 1973, 1974). As we have seen, stereopsis occurs when there is a partial failure in the fusion of the non-identical visual patterns fed into the two eyes. This disparity is resolved by the perceptual illusion of depth in accord with the learned interpretation of the coded information provided by the disparity detector neurones. Attention is required for this resolution of even the simplest random dot stereogram. With the very sophisticated stereograms designed by Julesz (1971) it takes a concentrated mental effort in order gradually to resolve the observed disparity into some quite shapely vision with the accompanying ecstacy of accomplishment! Similarly a skilled observer experiences delight when contemplating a masterpiece of painting in which the artist has depicted so well his experience of some scene or subject. All of this goes to show that our conscious selves are very fully engaged in effecting the transmutation from the encoded responses of cortical modules to the visually experienced pictures. Always the mental struggle is for meaning. The explanatory power of dualist-interactionism stands revealed. And conversely the frustration of trying to visualize the impossible figures (cf. Fig. 3-14) in three dimensions gives evidence of the commitment of our conscious self to achieve a meaningful resolution.

I realize that my explanation will be critized by Gregory, who would identify the postulate of readout by the conscious mind as the equivalent of an internal eye to see the trace and which 'would generate an endless regress of traces and eyes ... leading to an infinite distance with nothing different from the start of the regress' (Gregory, 1973). To which I would reply that the conscious mind must not be analogized to an internal eye, but rather it is the final stage of the perceptual process. There is, however, a way back to the neural events in the brain (cf. Fig. 1-7). For example, by attention (Lecture 4) we can amplify these events to give more effective presentation of the encoded binocular disparity in order to advance the resolution of a difficult stereopsis. As already mentioned, a strong mental effort is essential in arriving at a resolution. Also Julesz (1971) will be critical as he is very wary of the idea of having a homunculus or active perceiving agent in the head 'that sees the same things that we do'. He adopts the identity theory and anticipates the time when the feature extractors for the cyclopean vision of stereopsis 'are actually found and localized by neurophysiologists and neuroanatomists'. The philosophical position of the identity theory has already been criticized in Lecture 1. And of course 'the homunculus' is our self-conscious mind that naturally 'sees the same things that we do!' So the last word is with the dualist-interactionist interpretations.

Lecture 4

Electrical Responses of the Brain

Résumé

A general account is given of electrical responses generated by the brain and led from the human scalp, the electroencephalograms (EEGs). Their modifications under a variety of experimental conditions lead to explanations of their mode of generation by the neocortex in response to thalamo-cortical interactions.

Special kinds of EEGs, the *event-related-potentials*, are described in response to peripheral nerve or skin stimulation.

Electroencephalographic methods with advanced averaging techniques have been used especially by Desmedt to display the late phases of event-related-potentials that are induced by *attention* and that give perceptual experiences. In that way new insights have been obtained on the influence that the mental process of attention exerts on the neural machinery of the brain. It is suggested that this influence is selective in causing the development of spatio-temporal patterns of modular activity that result in conscious experiences.

Another mental process, *intention*, is responsible for the planning and carrying out of motor actions. The associated electroencephalograms, *the readiness potentials*, are described in detail both with respect to their location and their time course.

Finally these electrical responses are discussed in relation to their generation by the cerebral cortex under the influences of intention and attention. It is argued that the readiness potential of the Kornhuber experiments and the related Jung experiments give an empirical demonstration of the action of conscious willing on the neural machinery of the brain. This suggestion is comparable with that already made for Desmedt's experiments on the late event-related-potentials developed under the influence of attention. The general basis of these explanations lies in the hypothesis of dualist-interactionism, and not in the materialist theories of the mind, for according to these theories mental events *per se* cannot *effectively* act upon brain events (cf. Lecture 1).

4.1 Human Electroencephalogram (cf. Jung, 1963)

The enormous complexity of the brain, especially of the neocortex, makes the analysis of the electrical responses recorded from the surface of the scalp a formidable task. There had been earlier reports, but our present understanding dates from Hans Berger, who from 1929 onwards published a series of papers describing the very low voltage rhythmic responses at about 10/s, (the alpha rhythm) and also a faster rhythm at about 20/s (the beta rhythm). This pioneering work was neglected for some years, but confirmation by Adrian and Matthews (1934) began the tremendous development of electroencephalography as a clinical tool. It could reveal so much of the disordered brain activity, e.g. the wide variety of seizure phenomena characteristic of epilepsy. Nevertheless the mode of generation of these cortical potentials and the factors controlling their rhythmicity are still only partly understood.

It seems inconceivable that the electrical responses of the billions of neurones linked together in unimaginably complex patterns could result in the simple wave form of the alpha rhythm (cf. Fig. 4-1 A). It is the laminated arrangement (Figs. 2-2 and 2-6) of the sheet of cerebral cortex that gives it the means of generating electrical field potentials that can be picked up from the surface of the scalp. We can think of the cortex as being effectively an electrical dipole (Jung, 1963).

For optimum conditions in the generating of 'brain waves', as they are called, the subject must be fully relaxed with eyes closed. The alpha rhythm recorded from the scalp over the occipital lobe (cf. Fig. 1-1) is shown in the R and L traces of Fig. 4-1 A. Usually the voltage is no more than 50 to 100 μV, so stringent precautions must be taken to eliminate artefactual contamination. The parietal and posterior temporal lobes (cf. Fig. 1-1) also are effective generators of the alpha rhythm. More anterior regions of the brain usually deliver a small irregular beta rhythm at about 20/s, but the frontal lobe has some irregular alpha output.

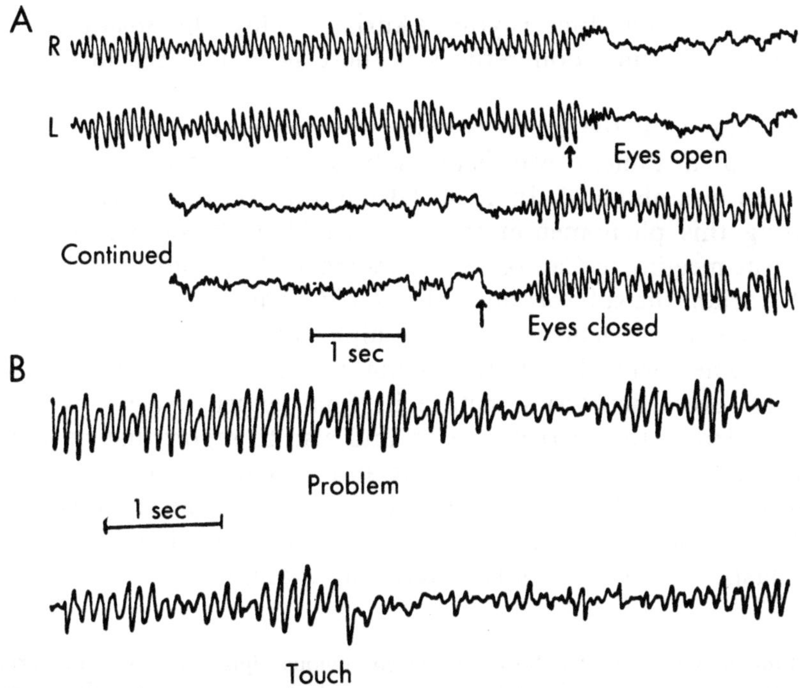

Fig. 4-1 A, B. Electroencephalograms of man. **A** Effect of opening eyes in suppressing the alpha rhythm recorded simultaneously from the right (*R*) and left (*L*) occipital poles of the human cerebrum, and return of rhythm on closing eyes. Note similarities of electroencephalogram on the two sides. Penfield, W., Jasper, H.: Epilepsy and the functional anatomy of the human brain. pp. 1–896. Boston: Little Brown & Co 1954. **B** Effect of brain activity in mental arithmetic and with afferent stimulation in causing temporary cessation of alpha rhythm. Adrian, E.D., Matthews, B.H.C.: The Berger rhythm: Potential changes from the occipital lobes of man. Brain *57*, 355–384 (1934)

In general we can regard these brain waves as being generated by the brain when there is minimal mental activity. Some years ago it became quite a fad to try to get into states for maximization of alpha rhythm – a harmless fun game. Figure 4-1 A shows that opening the eyes immediately abolishes the alpha rhythm. This does not occur if the visual field is entirely featureless and the subject does not 'try to see'. Trying to see also abolishes the rhythm when the eyes are opened in a completely darkened room. It seems therefore that the mental effort of attention is the key factor, not the visual stimulus *per se*. It is of interest that the brain of a narcotized animal also produces an alpha rhythm (Adrian, 1946). On closing the eyes the alpha rhythm soon returns if there is mental relaxation (Fig. 4-1 A).

In Fig. 4-1 B the alpha rhythm was depressed when a problem in mental arithmetic was being solved, and it returned when the solution was obtained. In 1934 Adrian and Matthews were demonstrating this phenomenon to the assembled Physiological Society that was meeting in Cambridge. A member of the audience suggested that Adrian (the subject) multiply 17 × 26. There was a rather longer and more severe depression of the alpha rhythm than in Fig. 4-1 B, then momentarily the alpha rhythm reappeared only to pass into further severe depression before its final return. Adrian was then able to deliver the correct answer, and also to report that he discovered that his first answer was wrong and he had to repeat the calculation more carefully. It was a striking display of the fragility of the alpha rhythm when confronted by mental effort. I quote Penfield and Jasper (1954) in another interesting example:

> Einstein was found to show a fairly continuous alpha rhythm while carrying out rather intricate mathematical operations, which, however were fairly automatic for him. Suddenly his alpha waves dropped out and he appeared restless. When asked if there was anything wrong, he replied he had found a mistake in the calculations he had made the day before. He asked to telephone Princeton immediately.

Figure 4-1 B also shows the effect of a touch in depressing the alpha rhythm obtained with eyes closed. The effectiveness of the touch depends on its unexpectedness, particularly as regards location, e.g. on the face. The 'startle value' is the significant variable, not the intensity of the contact. The same stimulus repeated will lose its startle value and hence its depressive effect on the alpha rhythm. In Fig. 4-2 A the eyes were open with attention to the visual field so the alpha rhythm was suppressed. At the signal the subject tried to hear the tick of an almost inaudible watch; attention was diverted from the visual field and the alpha rhythm reappeared for a second or so at full size and later at a depressed level until the watch was removed, when vision again completely suppressed the alpha rhythm. Since the recording was from the occipital lobe, the hearing input would have no direct influence.

Adrian (1946) sums up the alpha rhythm story very well:

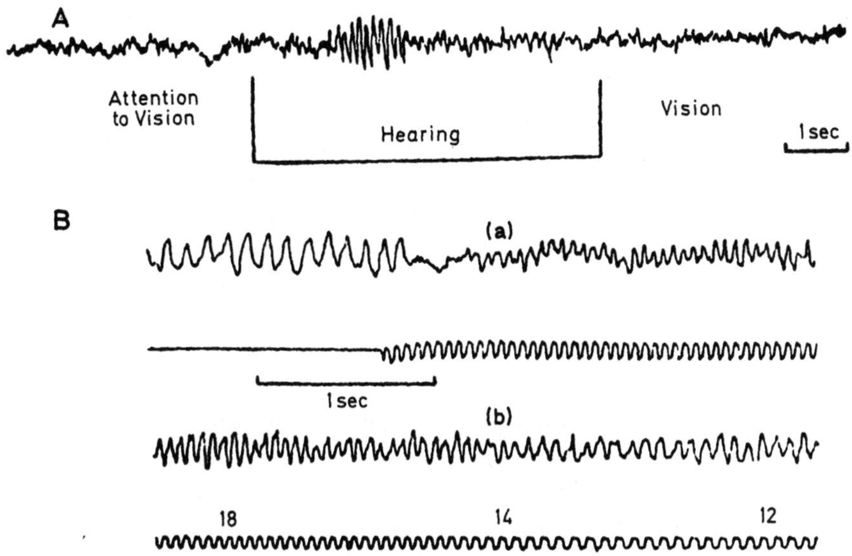

Fig. 4-2 A, B. Electroencephalograms of man. **A** Appearance of the rhythm when the attention is transferred from vision to hearing. The visual field has been made unattractive by 10D spectacles. During the middle section of the record the attention is concentrated on the tick of a watch and the alpha waves appear. **B** Records from the occipital region of the head, showing the effect of flickering light. Below the records from the head are tracings from a photoelectric cell showing the rate of flicker. In the lower record the rate was reduced from 18 to 12 per s, and the cerebral waves kept in step with the flicker. The cerebral waves are potential changes of the order of 20–30 μV. Adrian, E.D.: The physical background of perception, pp. 95. Oxford: Clarendon 1946

The α rhythm may be regarded therefore, as associated with the inattentive state and as occurring in regions of the brain which are playing no part in mental activity because the attention is directed elsewhere. With normal vision there is always something to catch the eye even though our attention is mainly elsewhere, and so the α rhythm is normally absent when we are awake. But as we become sleepy it is harder to keep the attention fixed, and the rhythm begins to reassert itself even with the eyes open; and when we begin to fall asleep it dominates the brain for a time, to be replaced later by occasional slow waves when the sleep is established (cf. upper traces of Fig. 6-8).

An ordinary visual experience causes the alpha rhythm to be replaced by a very fine irregular rhythm, which we can regard as being generated by the complex operations of the neural machinery in the visual processes (Adrian, 1946). Related to this explanation is the finding that, if the visual input on opening the eyes is a flickering light, there will be over a considerable frequency range a production of rhythmic waves by the occipital cortex in phase with the flicker. For example in Fig. 4-2B (a) there is an initial alpha rhythm which is momentarily suppressed by the flickering light at 18/s but soon beats in time with it. In Fig. 4-2B (b) the cortical rhythm is seen to follow the flicker from 18 to 14 to 12/s. Each flash of the flickering light is evoking an electrical potential in the occipital cortex in the manner described below in Sect. 4.2.

There have been many suggestions that the alpha rhythm is generated by cortico-thalamic reverberating circuits with experimental support from studies of thalamic repetitive activity. An excellent survey of this literature is given by Andersen and Andersson (1968) who report a more recent (1962 onwards) analytical study of the thalamic responses and of the simultaneous cortical potentials. Repetitive responses resembling the alpha rhythm can be produced in animals by barbiturate anaesthesia, an effect of great experimental utility. Figure 4-3 shows in the right column the relationship of barbiturate thalamic spindles A, B, C to the cortical spindles (a, b, c). There is much evidence that the generation of the thalamic spindles is due to the operation of circuits shown in Fig. 4-3 in the thalamus. The thalamic relay cells activate inhibitory interneurones by axon collaterals, thus each thalamic discharge to the cortex generates a feedback inhibition to the adjacent population of thalamic relay cells, which are thus brought into phase. A more widespread influence is exerted by the inhibition of the distributor cells that have a background excitatory action. In that way an immense number of thalamic relay cells will have synchronized inhibitions that have a duration of about 100 ms, terminating in a rebound discharge of impulses with another inhibitory action and subsequent rebound, and so on, as indicated in traces A, B and C of Fig. 4-3. Almost all inputs from lower levels of the central nervous system reach the neocortex via thalamo-cortical pathways, which are widely distributed to all parts of cortex (Fig. 5-3). The paths to three columns a, b and c are shown in Fig. 4-3 together with the potentials a, b and c generated in these columns. It will be appreciated that this hypothesis gives the prime role in spindle generation to the thalamic circuitry, the cortical columns merely responding to the repetitive thalamic drive by afferent fibres such as those illustrated, SPEC. AFF, in Figs. 2-2 and 2-6. However the recent concepts of modular operation of the

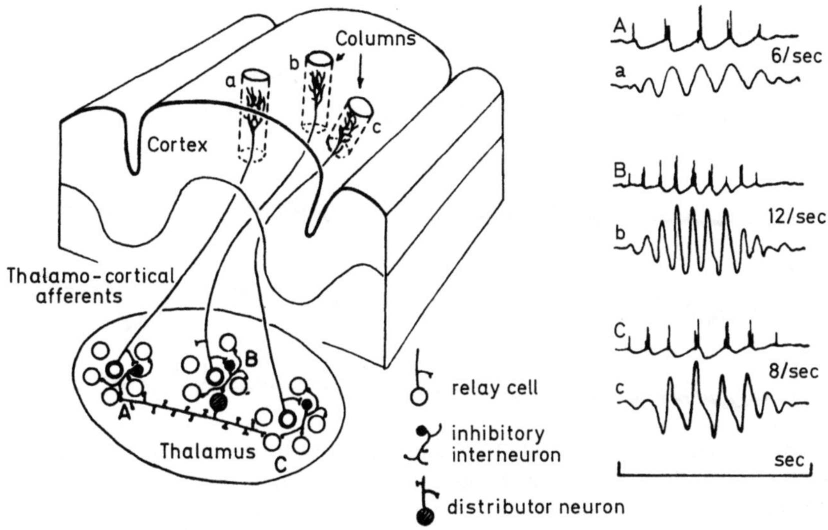

Fig. 4-3. Diagram illustrating a model of the thalamo-cortical rhythmic correspondence. The neurones of three thalamic nuclear groups, *A*, *B* and *C*, send their axons to the appropriate part of the cerebral cortex, activating the columns *a*, *b* and *c*. Collaterals of these axons excite inhibitory interneurones (*black*), which have profusely ramifying axons that can initiate post-synaptic inhibition simultaneously in a large number of thalamic neurones. The different thalamic nuclear groups have been given different intraspindle frequencies, times of onset and of stop. The corresponding features of the cortical spindles vary accordingly, as illustrated in the right hand column. The upper lines are imaginary spindles from the thalamic groups *A*, *B* and *C*, whereas the lower lines show the corresponding cortical spindles as they would appear at points *a*, *b* and *c*, respectively. Andersen, P., Andersson, S.A.: Physiological basis of the alpha rhythm New York: Appleton-Century-Crofts 1968

neocortex with the possibility of re-entrant loops makes it likely that the neocortex may contribute actively to setting the rhythmic activity that is expressed as the alpha rhythm. The cortical activity asserts itself when aroused by afferent inputs or by mental activity as indicated in Figs. 4-1 and 4-2B.

4.2 Event Related Potentials

4.2.1 Evoked Potentials in the Neocortex

A single electrical stimulus to an afferent nerve produces on the surface of the cerebral cortex (Fig. 4-4) or of the overlying scalp

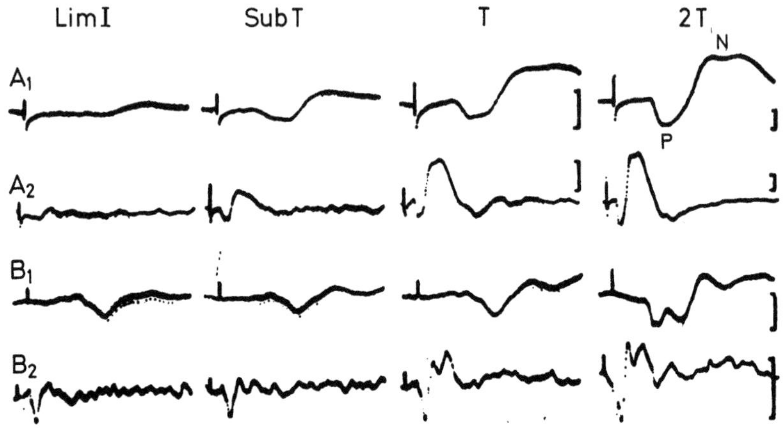

Fig. 4-4. Averaged evoked potentials of somatosensory cortex in relation to threshold stimuli at skin. Each tracing is the average of 500 responses at 1.8/s. Total trace length is 125 ms in A_1 and B_1 and 500 ms in A_2 and B_2; beginning of stimulus artefact has been made visible near start of each tracing. *A* and *B*, separate subjects, both parkinsonian patients. Vertical column *T*: threshold stimuli, subjects reporting not feeling some of the 500 stimuli. Column *2T:* stimuli at twice threshold current; all stimuli felt distinctly. Column *SubT:* subthreshold stimuli, none felt by subject; current about 15% below T in subject *A*, 25% below T in *B*. Column *LimI*: subthreshold stimuli at 'liminal intensity' (see text), about 25% below T in subject *A*, about 35% to 40% below T in *B*. Polarity, positive downward in all figures. Vertical bars in *A*, under T, indicate 50 μV in A_1 and A_2 respectively, but gains are different in 2T as shown; for B_1 and B_2, 20 μV bars. Libet, B.: Electrical stimulation of cortex in human subjects, and conscious sensory aspects. In: Handbook of sensory physiology. Iggo, A. (ed.), Vol. II, pp. 743–790. Berlin, Heidelberg, New York: Springer 1973

(Figs. 4-5 and 4-6) an initial positive (P) and later negative (N) wave. These evoked potentials, as they are called, have been extensively investigated with stimulation of afferent nerves of various modalities. A compilation of the earlier literature is given in a review (Eccles, 1951 a). It is generally agreed that the initial positivity is generated when the afferent impulses synaptically excite the neuronal dendrites in lamina IV (cf. the terminals of the AFF fibres in Fig. 2-2). The surface positivity is due to the sources of current flow from the superficial dendritic zones to the dendritic negativity in lamina IV (cf. Fig. 4, Eccles, 1951 a). The later surface negative potential would be generated by secondarily excited neurones.

However this initial diphasic wave of the evoked potential is not the neural counterpart of a conscious sensory experience. For example it is relatively unaffected in anaesthesia (Jasper, 1966). Also in Fig. 4-4 stimuli giving the Sub T responses were not perceived by the subject, the much larger T responses being at the perceptual threshold. However repetitive stimulation at 60/s for a fraction of a second did elicit a conscious experience even when at the weak Lim I level (Libet, 1973). Thus it can be concluded that for a conscious experience there must be an input to the somaesthetic cortex by many afferent fibres acting together or by repetitive action of a fewer number. The important point is that it requires a considerable build up of neuronal responses before the threshold for a conscious experience is reached. At the 2T level of the B_2 row of Fig. 4-4, later wave forms appear which signal further spread of neuronal activity in the cortex. Such late waves are abolished when anaesthesia eliminates sensory experience.

The wide dispersion over the cortex of the sensory input from touch is indicated by the depression of the alpha rhythm for some seconds (Fig. 4-1 B). The recording of cerebral circulation by the radio-Xenon technique as described in Lecture 6 (Ingvar, 1975a; Lassen et al., 1978) provides a more striking demonstration of the widespread cerebral activity resulting from a touch stimulus. Comparison of the 'Rest' with 'Touch' pictures in Fig. 6-2 reveals that with a continued touch on the hand there was a considerable increase in the cerebral circulation above the resting value in the frontal and parietal lobes. Thus the increased cerebral activity occupied an area very much as indicated in Figs. 3-1 and 3-2A from areas 3, 1, 2 to areas 5, 4, 6, 7 SM and 46. This wide dispersal would be along the module to module transmission lines indicated diagrammatically in Figs. 2-5 and 2-7. It can be conjectured that in Libet's experiments (1973) the subject does not have a conscious experience until there has been this widespread dispersal with the generation of complex spatio-temporal patterns as in Fig. 2-10. We shall now see that conscious attention greatly aids in this development.

4.2.2 Attention

4.2.2.1 Introduction

Adrian (1946) expresses the importance of attention in our perceptions to which brief reference was made in Lecture 3:

> The brain may be wide awake, but if we are looking at something interesting we may be quite unaware of noises which are certainly sending messages into the auditory area. To find out what happens when the attention is directed first to one field and then to another we must rely mainly on the human subject who can direct his attention more or less at will.

When trying this out for oneself it is best to try somaesthetic or auditory sensing because with vision the direction of vision dominates the direction of visual attention. With eyes shut so as to abolish the dominant visual input it is possible to concentrate attention on a small somaesthetic area, such as the left little finger or the right big toe and to discover some sensory experience to the virtual exclusion of all other perceptions. Likewise with hearing we can concentrate our attention on some chosen input, for example speech, to the exclusion of all others. As Adrian (1946) states:

> The direction of our attention seems to be determined also by a balancing of claims which must take place in some central region to decide which part of the cortex shall be set free from the α rhythm for the use of the mind. But how the direction of attention is decided and where it is decided is quite uncertain, and for the present we can only say that a condition for its operation is the suppression of the uniform beat of the α rhythm.

It should be noted that Adrian makes a clear reference to the liaison brain, and also speaks in a dualist-interactionist manner.

Mountcastle (1978a) stresses the importance of selection from the 'sea of signals' in which we are immersed. He emphasizes the role of the direction of gaze in providing selective visual attention, there being often sequential glimpses as the eyes move in rapid saccadic jumps from one gaze to another. But, when an object of interest is seen, there will be an attentive fixation upon it. Attention is effective not only in reading out from the selected modular operations, but also in contributing to their patterned development.

Rensch (1971) expresses well the nature of attention, writing dualistically:

> Our ideas do not normally succeed one another in a random course prompted simply by the existing trains of association. ... On the contrary, we pursue definite trains of ideas, that is, we direct our attention and guide our thoughts along certain lines by volitional processes.

4.2.2.2 Experimental Investigations on Attention

There have been many investigations on the effect of attention in modifying the evoked cortical potentials, particularly the potentials following the initial P and N waves (cf. Fig. 4-4). These later waves are of diverse polarity and latency and are assumed to be due to transmission from the initially excited focus into adjacent cortical areas and beyond. These later complex wave forms can now be lifted out of background noise by the technique of averaging. In that way potentials of a few microvolts can be recognized with assurance.

Desmedt and Robertson (1977a, 1977b) review a large number of investigations on the later complex waves that are generated by a brief somatosensory input. There is general agreement that under favourable conditions (Fig. 4-5B and C) a negative wave with a peak latency of about 150 ms (N 150) precedes a positive wave of peak latency 300 to 400 ms (P 300 or P 400). By careful experimental design they have been able to establish the conditions under which these waves are generated, and thus to throw light on the cerebral events that lead to conscious recognition of a touch.

In the standard experimental procedure the subject has bipolar stimulating electrodes fixed to the second and third fingers of each hand. The stimuli to each finger are brief electric pulses of about 3 mA, that are maintained at 30% to 50% above threshold throughout the experiment. The timing of the pulses is controlled by a random generator that delivers the pulses at an average frequency of 150/min, but the actual intervals vary from 250 to 570 ms. Furthermore the pulse applications to any one of the four fingers are also randomized so that any one finger will receive stimuli at an interval ranging from 250 to 2280 ms, the mean frequency being 37/min. In Figs. 4-5 and 4-6 the EOG traces show that in averaged records there were no significant eye movement artefacts. Also in Fig. 4-5D it is shown that, when the potentials generated by finger stimulation are cancelled by successive inversions, there is no residual potential due to background activity.

Figure 4-6 illustrates the experimental procedures. The subject is instructed to 'attend' to a particular finger. For example in Fig. 4-6B it is the second finger of the left hand in the left figurine and of the right hand in the right figurine, the attended fingers being shown in black. In Fig. 4-6B and C the electrocorticograms were

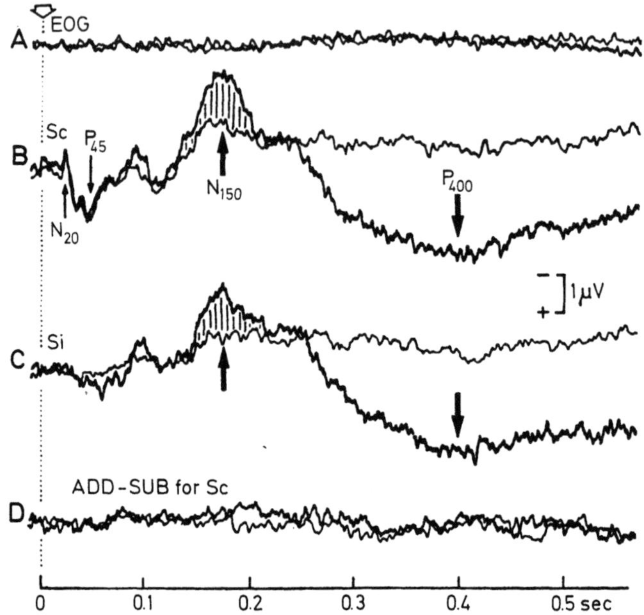

Fig. 4-5A-D. Cognitive SEP (somatosensory evoked potentials) components N 150 and P 400 at right and left parietal derivations. **A** The EOGs (electro-oculograms) document the absence of any remaining eye movement artefacts. **B, C** The SEPs were elicited by electrical stimuli of 3.5 mA to the third finger of the right hand in runs when the subject attends that finger (*thicker traces*) or in other runs when he attends acoustic clicks delivered through earphones (*thinner traces*). **B** Contralateral SEP, *Sc*, recorded from the parietal scalp derivation. **C** Ipsilateral SEP, *Si*, recorded from the symmetrical derivation. **D** Add-sub average checks for possible contribution of non-reponse EEG background to the averaged SEP response. The increase of N 150 to target stimuli (cross-hatched area) is somewhat larger contralaterally (**B**). The P 400 estimated from prestimulus baseline is virtually equal on both sides. Desmedt, J.E., Robertson, D.: Search for right hemisphere asymmetries in event-related potentials to somatosensory cueing signals. In: Language and hemispheric specialization in man: cerebral event-related potentials. Desmedt, J.E. (ed.), pp. 172–187. Basel: Karger 1977

averages of the responses evoked by those stimuli applied to the attended finger. During a run of many minutes the subject has to count the number of stimuli felt by the attended finger. At the mean frequency of 37/min this task fully engages the concentration of the subject. In a satisfactory run the subject will score about 95% of the correct number, as recorded by the computer. In Fig. 4-5B and C the background light trace was recorded when the subject was attending to acoustic clicks delivered by an earphone and not to the stimulated finger. Relative to these control responses there are in Fig. 4-5A large N 150 and P 400 waves, and in Fig. 4-6B large N 140 and P 400 waves. These waves stand out from the background records in lighter trace that are registered for the same finger stimulation, but when the subject is attending to acoustic clicks in Fig. 4-5B and C or to the same finger of the other hand, as indicated by the right figurine of Fig. 4-6B, and so was ignoring the stimuli to the finger that was giving the recorded electrocorticograms. The initial N 20 P 45 waves are superimposed, and the latency of origin of the later N wave varies from 50 to 130 ms in different subjects with a mean value of 77 ms. It should be noted that with scalp recording there may be a very small initial N wave (Fig. 4-5B) that was not seen with subdural recording in Fig. 4-4.

The same experimental procedures were applied in Figs. 4-5C and 4-6C, but recording was now over the ipsilateral parietal cortex. In view of the crossed representation of the fingers, the small initial P response was to be expected, but it was surprising to find a large late potential complex when the attended finger was stimulated. The N 140 or N 150 waves were smaller than for recording over the contralateral hemisphere, but the P 400 waves were approximately equal. Figure 4-6E and F provide an interesting variant in that there was the same attended finger, but the recording was from responses evoked by stimulation applied to the first finger either of the same (E) or of the opposite hand (F). Comparison of the thick traces with the background thin traces shows that with recording from either side there was a N 140 and a small later P wave. Thus the effect of attention to stimulation of a particular finger is largely localized to stimulation of that finger. In Fig. 4-5B there is accurate superposition of the two traces of the N 20 and P 45 waves, and this also occurs for the P 45 waves in Fig. 4-6B and E. This observation is of importance in indicating that attention has not influenced by 'gating' the transmission in the pathway from the periphery to the sensory cortex.

Desmedt and Robertson (1977a, 1977b) have found that it was important to have the attention of the subject fully occupied in order to secure large late N and P waves. They chose a frequency at the high rate of 150/min for this purpose. The late waves were much smaller at a frequency of 40/min for example. Also, if the stimuli were well above threshold, the attention of the subject was not fully occupied. The device of mental counting of the stimuli to the attended finger was adopted in order to eliminate complications by the muscle action potentials that would occur if any signalling system were used.

The subject was instructed at the outset to exclude from his mind all the non-target stimuli either to the adjacent finger or to the opposite hand. His attention was fully occupied in the counting of the target stimuli, so we can assume that the other stimuli were not consciously perceived. Thus these experiments lead to two conclusions:

(1) that by attending to a finger there is a remarkable increase in the spreading neuronal activity beyond the initial activity that its stimulation generates in the primary receiving cortex;

(2) that these later evolving responses are associated with the conscious perception of the stimuli. The effect of attention to a finger is not absolutely restricted to that finger, but to some extent influences the responses evoked by stimulation of the adjacent finger (Fig. 4-6E, F).

It is of very great interest that the spreading neuronal activity induced by attention eventually involves both hemispheres to an equal extent. For example a comparison of the N 140 or 150 waves in Figs. 4-5B and 4-6B with those from the other hemisphere in Figs. 4-5C and 4-6C shows that the contralateral response is considerably larger. On the other hand the P 400 responses are of the same size. In Fig. 4-7 there is plotting of a large number of IPSI and CONTRA responses. Included in the responses plotted in Fig. 4-7A are a number of P 100 responses (dots). For some unknown reason the initial attended response was in those experiments a positive wave with a maximum at about 100 ms instead of the N 140 wave. In Fig. 4-7A the N 140 responses are larger on the contralateral side, whereas in Fig. 4-7B the P 400 responses are very close to the 45° line. Thus the neuronal activity induced by attention is widely spread over both hemispheres. The recorded waves are largest over the parietal lobes, but are also of considerable size over the adjacent areas of the frontal lobes. Jones and Powell (1970b) have shown by degeneration techniques the sequential relays for neuronal transmission from the primary somaesthetic area (cf. Fig. 3-2). The spread is initially largely to Brodmann areas 5 and 7 (Fig. 3-1) in the superior parietal lobe, but later the Brodmann areas 6 and 46 in the frontal lobe are involved. Moreover all these parietal and frontal areas are

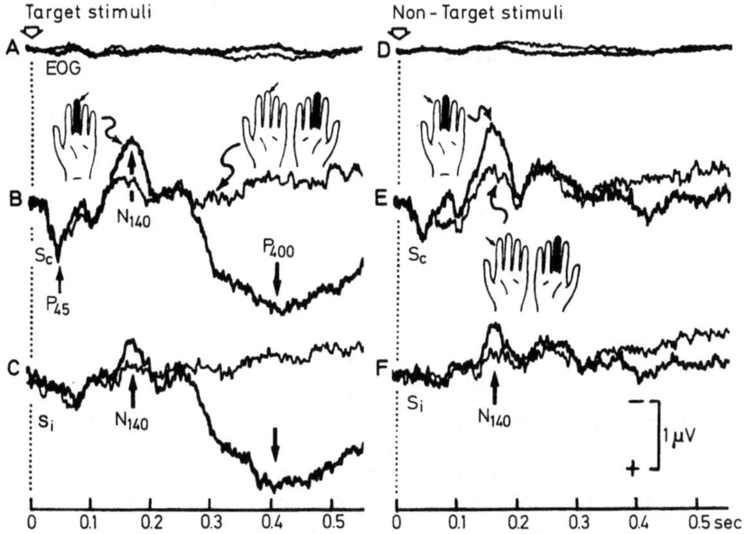

Fig. 4-6 A–F. Experiment with random sequence of stimuli to four fingers. **A** and **D**, EOG controls. **B** and **E**, SEPs recorded at the contralateral parietal focus (*Sc*) by stimuli of 3 mA to the third (**B**) or second (**E**) fingers of the left hand. **C** and **F**, corresponding potentials recorded simultaneously at the symmetrical ipsilateral (*Si*) parietal electrode. The subject counts the target stimuli to the third finger of either the left (*thicker traces*) or the right hand (*thinner traces*). In the hand figurines, the attended finger is represented in black and the *small arrow* points to the finger stimulus which evokes the SEP observed. N=560. Desmedt, J.E., Robertson, D.: Differential enhancement of early and late components of the cerebral somatosensory evoked potentials during forced-paced cognitive tasks in man. J. Physiol. *271*, 761–782 (1977)

very effectively connected to the contralateral side through the corpus callosum.

The effect of attention in developing a late positive wave is illustrated in Fig. 4-8 for a quite different task (Desmedt, 1977; Desmedt and Robertson, 1977b). The right index finger was dropped by an electro-mechanical device onto a circular ridge on the end of a Plexiglas rod. The test stimuli were provided by a gap in the ridge that could be oriented for successive tests in four directions (90° apart) relative to the body of the subject. In the control stimuli there was no gap in the ring. Figure 4-8 illustrates a typical experiment. The 'active touch' as signalled by the vertical dotted line leads to an

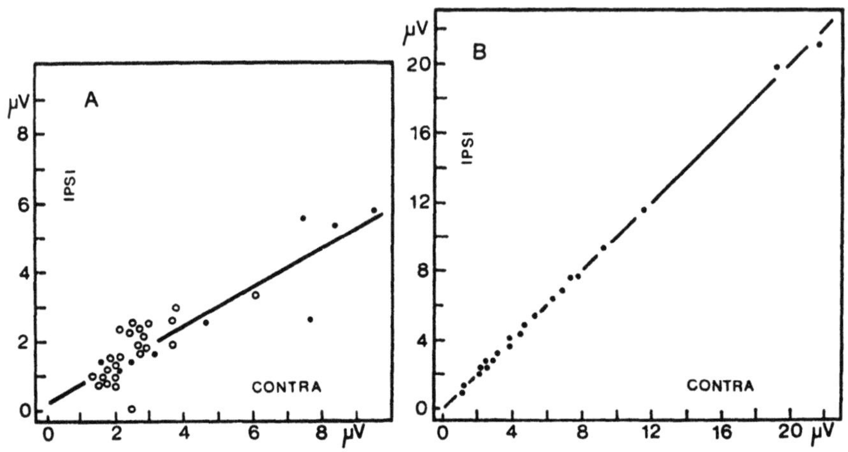

Fig. 4-7 A, B. Lateral distribution of SEP components in fast four-finger paradigms. Comparison of the peak voltages of the cognitive components N 140 (*circles* in **A**), P 100 (*dots* in **A**) and P 400 (*dots* in **B**) at symmetrical parietal derivations referred to the earlobe on the same side. The voltages at the contralateral (abscissa) and ipsilateral (ordinate) parietal locations are compared. The voltages in μV are estimated from the prestimulus baseline. The SEP components P 100 and N 140 are larger contralaterally to the attended hand, no matter whether the target stimuli are delivered to the right or left hand. The P 400 components are equal on both sides. Desmedt, J.E., Robertson, D.: Search for right hemisphere asymmetries in event-related potentials to somatosensory cueing signals. In: Language and hemispheric specialization in man: cerebral event-related potentials. Desmedt, J.E. (ed.), pp. 172–187. Basel: Karger 1977

initial positive and large negative wave, which are the evoked potentials (cf. Fig. 4-4). Then in B there is a prolonged positive potential (up to 2 s) much larger over the right than over the left parietal lobe. As in Fig. 4-5A and D and Fig. 4-6A and D the EOG and the ADD-SUB controls establish that the observed potentials in B, C and E are genuinely produced by the finger contact. It is most important that over the left parietal lobe (C) there was no significant difference between the smooth and ridge contacts, whereas in E over the right parietal lobe the long positive potential distinguished the ridge from the smooth contact. A similar late long positivity was also observed over the right hemisphere when the active touch was applied to the left hand. This special electrical response of the right hemisphere is in accord with the attribution of spatial

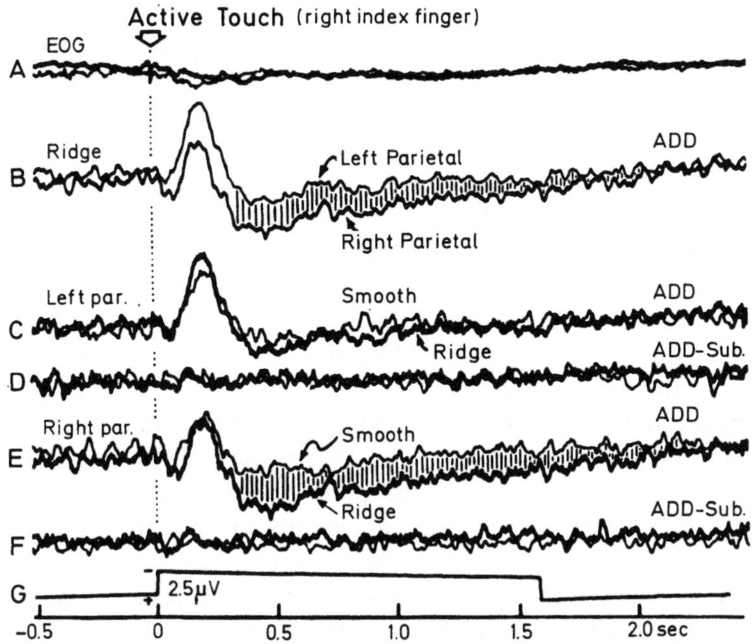

Fig. 4-8 A–G. Active touch by the right index finger in a right-handed subject with language located in the left hemisphere. The finger is dropped onto the Plexiglas ridge at the time indicated by the *arrow*. The traces represent the averages of 101 trial samples. **A** Electro-oculogram control over the same trials. **B** Palpation of the ridge; superimposed traces simultaneously recorded from the right (*thicker trace*) and the left parietal cortex. The vertical hatching indicates the extent of the positive electrogenesis on the right side. **C** Traces recorded over the left parietal cortex in two different sets of runs involving active touch of either the ridge (*thicker traces*) or a smooth Plexiglas surface (*thinner traces*). **D** Presents, as a control for non-response background activity, the add-subtract averaging of the same trial samples as in **C**. **E** Traces recorded over the right cortex in the same two sets of runs, thus involving active touch of either the ridge (*thicker traces*) or a smooth surface. **F** is the add-subtract average control of **E**. **G** Calibrating step function (2.5 µV). Negativity of the active electrode records upward. The reference electrode is at the midline vertex, which minimizes interferences from muscle potentials and from alpha waves of the electroencephalogram. Similar features were recorded with earlobe references in this experiment. Desmedt, J.E.: Active touch exploration of extrapersonal space elicits specific electrogenesis in the right cerebral hemisphere of intact right-handed man. Proc. Natl. Acad. Sci. USA *74*, 4037–4040 (1977)

discrimination to the right hemisphere of right handed subjects, as described in Lecture 1 (Fig. 1-5). Furthermore this late response arises because the attention of the subject is directed to the task of identifying the gap and its orientation.

4.2.3 Influence of Attention and Intention on the Neural Machinery of the Brain

Let us consider the sequence of events when attention is effective in recognition of a weak finger stimulus that otherwise goes unnoticed, as in the experiments of Figs. 4-5 and 4-6. In both cases there is the same afferent input to the somatosensory cortex, as is shown by the identical N 20 and P 45 waves. Under the influence of attention that fully engages the subject, there is an extensive development of neuronal activity which spreads widely on the ipsilateral side and then progressively invades the contralateral side so that by 400 ms there is symmetrical involvement of both sides (cf. Fig. 4-7B). I would propose that only in these latter stages of widespread neuronal activity is there the readout by the mind that gives the conscious experience of the finger stimulus which is mentally counted. This long latency of a conscious recognition matches the experimental measurements of Libet (1973) using weak repetitive stimulation of the somaesthetic cortex of the conscious human subject.

It is now possible to give a more detailed explanation of the mechanism of spreading activation of the neocortex. As described in Lecture 2 it has been established that the association cortex is composed of modules that are about 250 μm across and that function as units both in receiving input (cf. Fig. 2-7) and in projecting ipsilaterally and contralaterally to other modules (Fig. 2-5) (Goldman and Nauta, 1977; Szentágothai, 1978a). So we have to envisage that under the influence of attention there has been a great development of a spatio-temporal pattern of modular activities. It is this pattern which eventually is detected by the conscious mind in the process of attention. This is shown diagrammatically in Fig. 1-7, for communication from the activated modules of the liaison brain to the conscious mind. But, furthermore, in Fig. 1-7 there is shown communication from the mind to the modules of the brain. Presumably this is the channel of activity for the influence of this conscious process of attention whereby it effects the development of neuronal activity from the initial input to the somatosensory cortex. The process of attention illustrates both of the channels of communication shown in Fig. 1-7. Thus it provides important empirical evidence for the dualist-interactionist hypothesis of the relationship of the mind to the brain (Lecture 2 and Popper and Eccles, 1977).

The potentials related to the recognition of the ring-gap and its orientation similarly involve the spread of neuronal activity from the initial input to the somatosensory cortex. But in this case the mental effort to recognize and judge the orientation of the gap results in the modular spread being diverted to the right hemisphere where there is a special aptitude for spatial judgement in right handed subjects (Fig. 1-3). So we are introduced to the concept that the development of the modular pattern of neuronal activity in the neocortex is dependent not only on the attention of the subject, but also on the interest in extracting meaning from the sensory input. Desmedt and Robertson (1977a and 1977b) have opened up a rich field of investigation on the brain–mind problem.

In conclusion it should be stated that, important as these potentials are as signals of neuronal activity in the neocortex, there is still no information on the detailed mode of the generation of the negative and positive potentials obtained by scalp leads.

4.2.4 Intention

4.2.4.1 Introduction

With his superb insight Sherrington (1940) provides the opening theme:

> The climax of mental integration would seem to be 'attention'. ... The 'willed' act is but a culmination of attention. Where is an instance of completer solidarity in a complex organism than a man intent and concentrated upon some act of strenuous 'will'? — The anthropoid primates have not it. But can we imagine man, even primitive man, without it? Rodin's statue portrays prehistoric man erect with hand to forehead essaying abstract thought. His too is that other statue, seated, the 'penseur' absorbed in abstraction.

Beloff (1976) expresses very well the essential nature of intention:

> A distinguishing attribute of mind ... is the familiar intentional or purposive aspect of behaviour which transforms what otherwise would be a mere sequence

of movements into a meaningful action. A machine can go through any sequence of movements which the ingenuity of its inventory will allow but no purely physical system can act in the sense in which this implies attention. ... To equate the cognitive processes of human beings or animals with the information processing of computers is to confuse that which is simulated with its simulation.

When considering the preprogramming of action Jung (1980) states:

By directing the gaze toward the anticipated object we use directed attention and perform a preparatory act of anticipation of a future goal. For both the attentively sensed percept and the voluntarily planned action we anticipate future events. The programs of anticipating purpose imply a reversal of time of the present and future which is unreal and can only exist in the mind. However, there is objective evidence that the brain is involved in the anticipation.

For example Mountcastle et al. (1975) and Mountcastle (1978a) show that neurones of Brodmann area 7 (Fig. 3-1) of monkeys have special roles in the tasks of fixation and tracking of observed targets (cf. Fig. 3-3). However, they have an anticipatory orientation because their activation only occurs when the monkey is motivated to direct attention, for example towards food when hungry or towards liquid when thirsty. In addition there are other neurones in area 7 also with an anticipatory role in that they are related to arm projection and to hand manipulation with the aim of reaching and investigating tempting objects within the extrapersonal space (Fig. 3-3).

Some general remarks by Thorpe (1966, 1974) can well conclude this introduction to attention:

Consciousness is a primary datum of existence and as such cannot be fully defined. The evidence suggests that at the lower levels of the evolutionary scale consciousness, if it exists, must be of a very generalized kind, so to say unstructured. And that with the development of purposive behaviour and a powerful faculty of attention, consciousness associated with expectation will become more and more vivid and precise.

4.2.4.2 Cerebral Correlates of Intention

There are three kinds of experimental investigations on the electrical potentials which are produced by the intention of movements and by the movement itself. All demonstrate the production of negative potentials of the cerebral cortex which can be recorded from the overlying scalp.

A) *Contingent negative variation (CNV)* was described by Walter et al. (1964). In the typical situation there is a period of expectation between the first (warning) stimulus and the second (imperative) stimulus some seconds later. The stimuli can be applied through the visual, auditory or somaesthetic senses. The CNV is really a contingent result of a Pavlovian conditioning, being appropriately called an expectancy wave, because it is generated by the expectancy of the imperative stimulus, the CNV beginning a second or more before it. The CNV is symmetrical and is optimally recorded from the scalp over the frontal cortex, being negligible over the parietal cortex. It is not dependent on a motor response to the imperative stimulus.

B) Secondly, there is the *readiness potential (RP)* which Kornhuber and Deecke discovered also in 1964. This preliminary account is designed just for comparison with the CNV. A much fuller treatment will follow because it is central to our theme of intention. The RP is the negative potential (Fig. 4-9) that is generated during the *intention* to carry out a *brief voluntary action*. It is most important to recognize that the RP differs from the CNV in that it is generated by a voluntarily intended action and not by a conditioning stimulus. Also it is optimally recorded over the parietal and precentral cortices (Fig. 4-9, 4-11 and 4-12) and not at all over the prefrontal lobe, where there is usually a small positive potential. It also differs from the CNV in that it is slightly asymmetric. Finally the CNV does not depend on the motor act whereas the RP can only be recorded if there is a motor act to trigger the averaging and processing computer. However Lassen et al. (1978) found out that, when the subject performed hand moving *mentally* without actually moving the hand, there was a pattern of blood flow increase over the precentral, the parietal and the supplementary motor cortex (Fig. 6-1). Such an experiment cannot of course be done for readiness potentials. The muscle action has to occur in order to trigger the backward computation of the computer.

C) The third condition producing surface negative potentials occurs with the voluntary control of the monitored goal-directed movement towards a target (Grünewald-Zuberbier et al., 1978). As seen in Fig. 4-10 it begins with a typical RP potential (in frame A), but the negativity over the parietal and precentral cortices is maintained

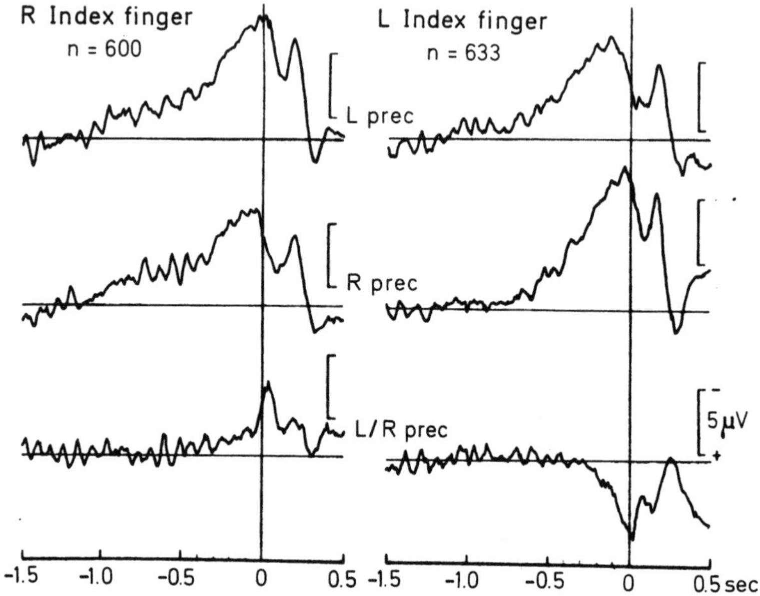

Fig. 4-9. Readiness potentials over the left or right parietal areas of the cortex in response to brief voluntary flexions of the right or left index finger as indicated. Zero time is at the onset of the electromyogram, the preceding potentials being derived by backward computation. 600 responses averaged for right finger, 633 for the left. Negativity preceding movement is larger in contralateral than in ipsilateral precentral leads. Thus, the bipolar recording left versus right precentral (*bottom graphs*) shows an upward deflection preceding right-sided movement, and a downward deflection preceding left-sided movement. With left-sided movement a particularly large right precentral RP is seen due to the left-handedness of this subject. Deecke, L., Grözinger, B., Kornhuber, H.H.: Voluntary finger movements in man. Cerebral potentials and theory. Biol. Cybern. *23*, 99–119 (1976)

at about twice the RP size during the movement (frame B) which may last several seconds and like the RP they are largest over the precentral and parietal cortices on both sides.

Jung (1980) has very well summarized the essential features of the three intentional processes.

> All these intentional movements and the accompanying cortical potentials are related to *attention*, that prepares A and B and controls C. Although a sensory evoked expectation of the second stimulus which elicits the motor response in the conditioning experiments (A) is different from a free decision to act (B), there are essential similarities. Both prepare action and correspond to an intention to act. In A during attentive expectation of the second stimulus which triggers

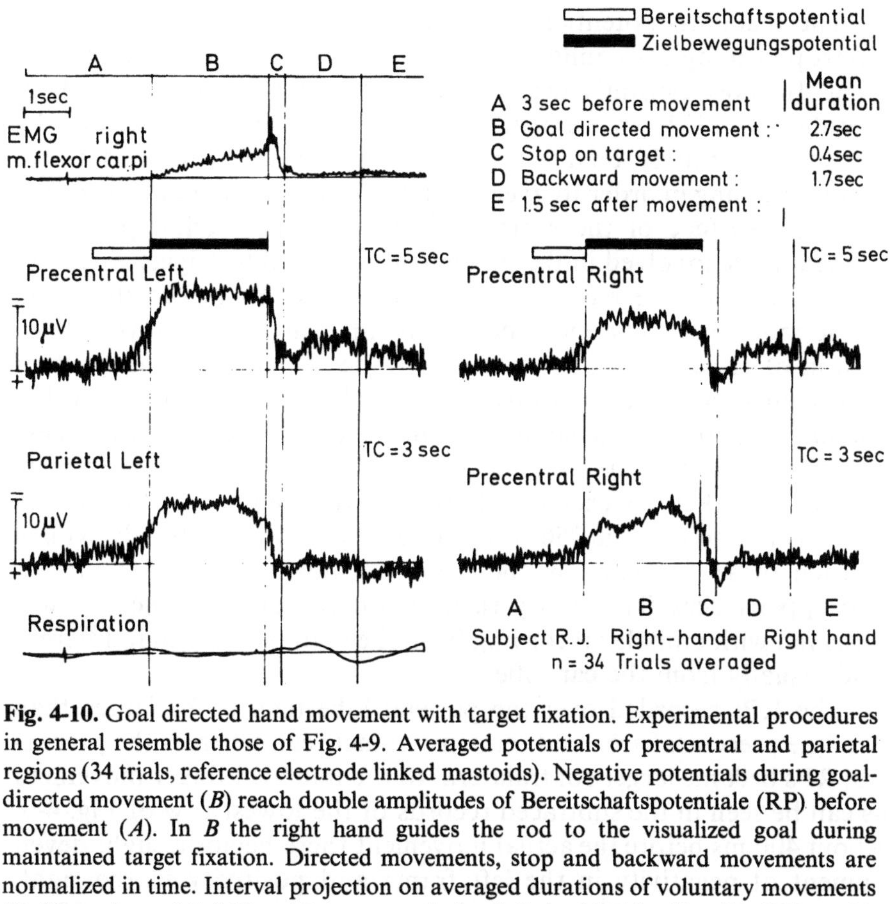

Fig. 4-10. Goal directed hand movement with target fixation. Experimental procedures in general resemble those of Fig. 4-9. Averaged potentials of precentral and parietal regions (34 trials, reference electrode linked mastoids). Negative potentials during goal-directed movement (*B*) reach double amplitudes of Bereitschaftspotentiale (RP) before movement (*A*). In *B* the right hand guides the rod to the visualized goal during maintained target fixation. Directed movements, stop and backward movements are normalized in time. Interval projection on averaged durations of voluntary movements (*B–D*) varying ±0.3, 0.08 and 0.4 s, respectively. Grünewald-Zuberbier, E., Grünewald, G., Jung, R.: Slow potentials of the human precentral and parietal cortex during goal-directed movements. (Zielbewegungspotentiale). J. Physiol. *284*, 181P–182P (1978)

a conditioned intended action, the attention is directed primarily to the outer world of the senses and secondarily to the intended movement. In B, before the freely decided movement, attention is internal and directed only toward the inner world of decision. The triggered action A differs from free action B mainly in its temporal determination by the sensory stimulus. Also during monitored action C the perceptual and voluntarily guided process of goal fixation and reaching is related to intention and attention. Hence, these surface negative cortical potentials may be considered as *cerebral correlates of intentional decision processes of willed action.* Of course, the neuronal correlates of these potentials are unknown.

The readiness potential is of particular interest because it is the electrophysiological counterpart of willing a free voluntary movement. In most elegantly designed experiments (Deecke et al., 1969; Kornhuber, 1974) an elementally simple movement is executed by the subject entirely on his own volition and yet there had to be accurate timing in order to average the very small potentials recorded from the surface of the scalp. The onset of the action potentials of the muscle involved in the movement was used to trigger a reverse computation of the recorded potentials up to 2 s before the onset of the movement. The movement illustrated in Fig. 4-9 was a rapid flexion of either the right or the left index finger, but many other movements have been investigated, even those of speech (cf. Eccles, 1979a, Fig. 10-4) with similar results. The subject initiates these movements 'at will' at irregular intervals of many seconds, extreme care being taken to exclude all triggering stimuli. In this way it was possible to average 250 or more records of the potentials evoked at each of several sites on the scalp, as shown in Fig. 4-9 for the two upper traces. The slowly rising negative potential, the RP, was observed with unipolar recording from the scalp against an indifferent lead, usually from the ear lobe.

The RPs recorded over the parietal lobe tend to be larger than over the precentral area and remain symmetrical, whereas the precentral RPs become larger contralateral to the intended finger movement as can be seen in the subtraced records of the lowest row of Fig. 4-9. About 400 ms before the actual movement there begins a slight development of negativity in the left frame and positivity in the right. Then this slow potential change is greatly increased at about 50 ms before the movement, being now strictly localized to the area over the cortical pyramidal cells responsible for conveying the central 'decision' to the motoneurones via the pyramidal tract (cf. the motor transmitting area of Fig. 1-1); hence the name motor potential (MP) for this wave. After the movement has been initiated, at the time of the vertical line in Fig. 4-9, there are complex potentials, which are partly due to muscle action potentials and which do not concern our study of intention.

When the movement is not of the rapid ballistic type, but is a smooth continuous action, the RP starts significantly earlier (Becker et al., 1976). In Fig. 4-11 the subjects pushed a rod along a tube at various speeds, but otherwise the experiment was carried out as in Fig. 4-9. In Fig. 4-11 the RP resembled that of Fig. 4-9 for

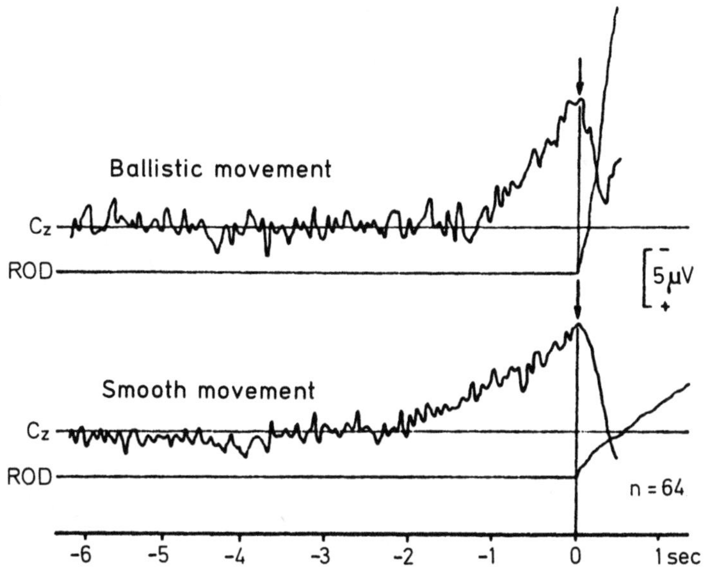

Fig. 4-11. Comparison of rapid and slow hand movements. The subject was pushing a rod into a tube at irregular intervals and varied the speed of the movement at random. RP preceding slow (smooth) movements (less than 10 cm/s) starts significantly earlier and is usually larger than RP preceding rapid (ballistic) movement (more than 100 cm/s). Becker, W., Iwase, K., Jürgens, R., Kornhuber, H.H.: Brain potentials preceding slow and rapid hand movements. In: The responsive brain. McCallum, Knott (eds.), pp. 99–102. Bristol: Wright 1976

the ballistic movement with an origin at about −1.0 s, but, when the movement was slow it started as soon as −2.0 s (−1.3 s on the average as against an average of −0.8 s for the ballistic).

An interesting pathological finding is indicated in Fig. 4-12 which is typical of many experiments on patients with unilateral parkinsonism. Such patients have severe movement difficulties on the side contralateral to the lesion of the basal ganglia and the associated motor cortex. Figure 4-12A shows a control series on a normal subject much as in Fig. 4-9. By contrast in Fig. 4-12B the left-sided patient with Parkinson's disease gave for right finger movement a fairly normal RP of the left precentral cortex but even a trace of a reversed RP from the right pathological cortex, hence the large subtracted negative potential in the lowest row. On the other hand the left paretic finger movements were associated with low RPs on both sides. Thus in hemiparkinsonism the RP was both unilaterally

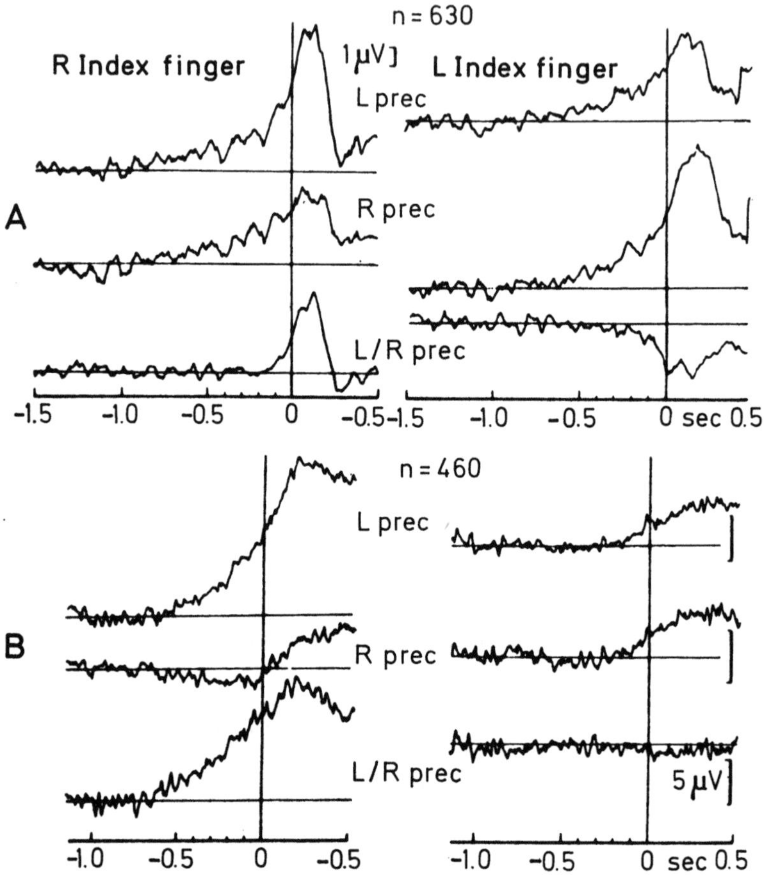

Fig. 4-12 A, B. Comparison of right and left-sided finger movements in a normal subject and in a hemiparkinsonian patient. **A** In the normal subject, negativity preceding movement is always larger in contralateral than in ipsilateral precentral leads (contralateral preponderance of negativity). Thus, bipolar recordings left versus right precentral (*bottom graphs*) are mirror images. Averages of 630 right-sided and 630 left-sided movements performed alternately in blocks of ca. 100 in the same experiment. **B** In a left-sided hemiparkinsonian patient, when moving the good (right) finger, the contralateral preponderance of negativity is grossly enhanced, but it is absent when moving the akinetic (left) finger. Thus, the downward deflection of **A** (*bottom right*) is lacking, because the intact although ipsilateral precentral region generates as much negativity as the contralateral affected precentral region. Deecke, L., Englitz, H.G., Kornhuber, H.H., Schmitt, G.: Cerebral potential preceding voluntary movement in patients with bilateral or unilateral Parkinson Akinesia. In: Attention, voluntary contraction and event-related cerebral potentials. Desmedt, J.E. (ed.), pp. 151–163. Basel: Karger 1977

reduced in amplitude over the affected cortex (Fig. 4-12B, left middle frame) and bilaterally reduced preceding movement of the akinetic finger (Fig. 4-12B, right frame). With bilateral parkinsonism all RPs were greatly reduced over the precentral cortex on both sides, but it was quite high over the midline vertex more anteriorly (Deecke and Kornhuber, 1978). Apparently the supplementary motor area was still receiving activation, possibly from the limbic system whereas the motor cortex suffered from lack of input from the destroyed basal ganglia.

4.2.4.3 Neural Mechanisms of Intention

These experiments at least provide a partial answer to the question: What is happening in my brain when an intended action is being programmed? It can be presumed that there is firstly a symmetrical activation of spatio-temporal patterns of modules in the parietal and precentral areas that, at about 400 ms before the willed movement, begin to converge on the opposite hemisphere (Figs. 4-9 and 4-12A). Finally, at about 50 ms before, the modular activity eventually comes to activate the pyramidal cells in the correct motor cortical area (Fig. 1-1) for bringing about the required movement. The manner in which modular activity spreads through the neocortex of both hemispheres to generate patterns has been illustrated in Figs. 2-5, 2-7, 2-8, 2-9 and 2-10 in very simplified diagrams.

In our present enquiry into the nature of intention it is important to recognize that, in contrast to reactions to sensory inputs, such as the CNV, the RP is generated by a mental event, an act of will. The experiments have been scrupulously designed to eliminate all other triggers. It is not proposed that the whole RP is generated by an act of will. Rather the RP is started in some small way by slightly changing the spatio-temporal patterns of modular activity and these build up by the connectivities shown diagrammatically in Figs. 2-5 and 2-7. This initial diffuse activity must be guided by the voluntary intention eventually to the correct motor pyramidal cells. During the long duration of the RP there will be activated circuits to preprogram the movement in all of its motor complexity both to the cerebellum (Allen and Tsukahara, 1974) and to the basal ganglia (McGeer et al., 1978, Chap. 13-5). The former circuit is revealed by the finding that in movements of primates there is activation of the cerebellar dentate nucleus before the motor pyramidal discharge (Thach, 1970). The latter circuit is revealed by the grave effects of basal ganglia disease on the RP (Fig. 4-12). The proposed effect of the willed *intention* on the responses of cortical modules is diagrammed in Fig. 1-7. Though this is an initiating and controlling influence, it need contribute only to a small degree to the build up of the RP and its eventual concentration onto the correct motor pyramidal cells in the MP response.

MacKay (1978) criticizes a similar statement that I wrote in a recent book (Popper and Eccles, 1977, p. 294) on the grounds that the RP could arise 'by influences entirely within the neuronal machinery' and that 'they do nothing to establish the existence of action across the interface between the ... mind ... and the modules of the cerebral cortex'. It is suggested that the backward average of the potential to give the RP before the movement could arise from 'some internally generated component of brain potential' and that the movement is triggered when this potential reaches some threshold value. This criterion for movement initiation overlooks the essential point of the Kornhuber experiment, namely that the subject is *consciously willing the movements* that give the muscle responses initiating the backward computation that reveals the RP. *The movements only occur on voluntary intention, as reported by the subjects.* It is the voluntary intention that I diagram in Fig. 1-7. The only alternative is to maintain that the subjects are illuded in their belief that they voluntarily initiate the movements, whereas the movements are *being entirely generated by the neuronal machinery.* This would be an explanation in accord with identity theory, but I do not think that MacKay would accept it.

In reference to intentional experiments and their relation to purpose and freedom of the will Jung (1979) states in personal communication:

> Human purpose and its relative freedom of will and action coexist with a causally determined world. This coexistence has developed through human civilization towards a diminished dependence of the mind on outer circumstances, because man has learned to use causal relations for goal-directed action and has systematized his knowledge in science and technology, i.e. in Popper's World 3. ... In Popper's terms: World 2 is only relatively free for willed acts when it uses World 3 information for cognition and action. The cognitional selection of percepts at higher conscious levels is intentional and guided by attention. Directed cognition is analogous to aimed action by using anticipation, purpose and memory.

This section on intention can very fittingly be concluded by Sherrington (1940, p. 325):

> The concept 'self' taken with all its connotations has become vastly far-reaching and intricate. Yet it would seem to have at its core an element relatively simple — the awareness or consciousness of each of us, prominent in certain of our motor acts, relates the self to the act. The awareness is of course an example of what in the abstract is spoken of as mind.

Sherrington goes on to point out that this intentional awareness is traceable to two sources: one is sensual with a detailed perception of the act derived from a battery of sense organs and having spatial reference, e.g. of the limb moving; the other is the direct awareness of the I-doing or of the self doing, and it is not derived from sense. This intentional action of the self is in remarkable accord with the hypothesis of dualist-interactionism. The proponents of the materialist theories of the mind have no explanation to offer.

Lecture 5

The Emotional Brain

Résumé

A wide variety of *affects* (the agreeable and the disagreeable subjective experiences) are distinguished from their expression in *emotions*.

The complex *limbic system* at the base of the brain plays a key role in all varieties of affects and in their emotional expression. After an introductory section on its anatomy and connectivities there is an account of experimental demonstrations of the association of emotional expressions with the limbic system and the related hypothalamus both in animals and in patients. Self-stimulating procedures in the brains both in animals and in patients distinguish between limbic-hypothalamic zones for pleasurable sensations and those for adversive reactions. For these polar opposites there are corresponding pharmacological properties: the catecholamines, noradrenaline and dopamine, for aggressive and feeding behaviour; serotonin for relaxation and sleep.

Since pain is of outstanding importance as an affect leading to emotional expression, there is an account of pain in its various manifestations. There are comprehensive descriptions of the anatomy and physiology of the nociceptive (pain) pathways from the dorsal root inputs up to the brain. A clear distinction is made between the *neospinothalamic tract* that mediates sharp-pricking pain and the *paleospinothalamic tract* that carries the nociceptor impulses leading to the slow-burning and the agonizing pain.

The pharmacological treatment of pain is discussed. It is recognized that even with moderate pain there is an intensive involvement of large areas of the neocortex, as shown by the considerable increase in blood flow in the frontal and anterior parietal areas of the brain.

The responses evoked by nociceptive input into the thalamus, the cerebral cortex and the limbic system are considered in relation to the tentative development of a general hypothesis of the cerebral events concerned in the generation of pain and suffering.

Finally there is a brief account of some other affects and their associated emotions.

5.1 Introduction

MacLean (1970) expresses very well the relationship of emotions to the brain and to the important concept of 'affect'.

> We commonly speak of the subjective and expressive aspects of emotion. As the subjective aspect is purely private, it must be distinguished by some such word as 'affect'. Only we as individuals can experience 'affects'. The public communication of affects requires expression through some form of verbal or other behavior. The behavioral expression of affect is appropriately denoted by Descartes' meaning of the word 'emotion'... Affects differ from other forms of psychic information insofar as they are subjectively qualified in a physical sense as being agreeable or disagreeable ... There are no neutral affects because, emotionally speaking, it is impossible to feel unemotionally.

In Table 5-1 the *affects* are listed as agreeable and disagreeable and furthermore, are categorized as basic, general and specific in accord with MacLean's suggestions. As he defines affects they are purely subjective, but normally are expressed as an appropriate emotion, for example by disgust for the items in disagreeable/specific. We differ greatly in the degree to which affects lead to emotional expressions. Characters can vary over the range from mercurial to stolid. In certain situations we may be able to stifle all emotional expression of an affect. Our literature is richly endowed with characters who dissemble so that their hate leads to tragedy of their unsuspecting victims. What better example can I give than Shakespeare's Iago? At the other extreme is the love that remains unexpressed often with mutual tragedy. And in between these two extremes there are all the ranges displayed in the characters of our literature. The literary artist is as it were painting a canvas with the affects of his characters as his ideas, and their interweaving emotional expressions as his story.

Much of Table 5-1 is for our everyday affects, or feelings as we commonly call them. There is no need to enlarge on them. If we are fortunate, they lead to enjoyments not only in the act, but also in the remembrances of past acts and in the anticipation of future acts. I am not advocating a hedonistic maximization of pleasures, nor even a maximization of happiness and satisfaction. That would give but a shallow life. We have also to seek for discontents and to be challenged by them, as we do for example by some unsolved problem. I can even be happy in such discontents because wrestling with a problem gives an anticipation of some achievement attained in a partial solution. That gives the essential motivation for scientific effort, namely to define problems so that their investigation may lead to a better understanding of some natural phenomenon.

Table 5-1. Table of Affects under headings, Agreeable and Disagreeable and in categories, Basic, General and Specific

	Agreeable	Disagreeable
Basic	Alleviation by eating drinking gratification	Deprivation of food (hunger) fluid (thirst) sex (longing)
General	Cordiality Friendliness Enthusiasm Love Worship Exaltation Ecstacy	Hate, hostility Anger, rage Violence Fear Guilt Anxiety Mental pain, anguish Depression Despair
Specific	Alleviation of Alleviation of Alleviation of Good odour Good taste Good sounds Good sights	Pain Nausea Vertigo Bad odour Bad taste Bad sounds Bad sights

[a] MacLean, P.D.: The triune brain, emotion, and scientific bias. In: The neurosciences. Second Study Program. Schmitt, F.O. (ed.), pp. 336–349. New York: Rockefeller University Press 1970.

5.2 Limbic System

Central to the brain structures concerned in emotional expression is a complex assemblage of nuclei and connecting pathways at the base of the brain that MacLean (1970) calls the limbic system. Experimental investigations of the emotional responses of the brain will be described after the most important features of the anatomy have been described.

5.2.1 Some Anatomical Considerations

In Fig. 5-1A the assemblage of nuclei (A) called the amygdala is the centre of the action. It is so named because of its almond-like shape. It is embedded on the inner side of the temporal pole beneath area 34 in Fig. 3-1. As shown by the arrows, it projects directly to the hypothalamus and also indirectly via the stria terminals. The hypothalamus is shown by vertical hatching (H). We will not be concerned with its many constituent nuclei. Brodal's (1969) subdivision of the amygdala into a cortico-medial and a baso-lateral group is valuable in that it accords with most studies on the effects of

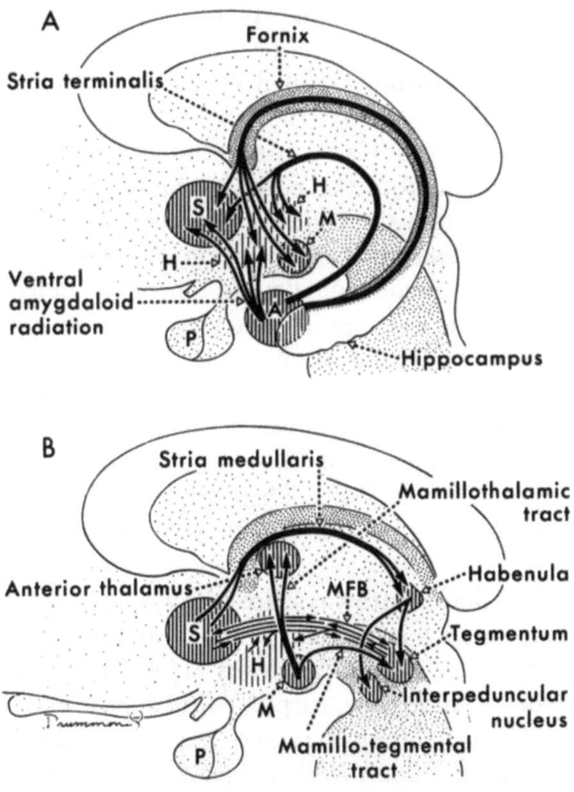

Fig. 5-1 A, B. Some anatomical interconnections between the limbic system and brain stem. Abbreviations: *H*, hypothalamus; *S*, septal area; *M*, mammillary body; *A*, amygdala; *P*, pituitary; *MFB*, medial forebrain bundle. **B** is sagittal section close to the midline; **A** is more oblique so as to include both the amygdala and the septum. McGeer, P.L., Eccles, J.C., McGeer, E.R.: Molecular neurobiology of the mammalian brain, pp. 644. New York: Plenum 1978

stimulation and ablation. Also shown is a path from the amygdala to the septal nuclei. Not shown in Fig. 5-1 A are the inputs to the amygdala from the olfactory bulb either directly or via the piriform cortex (cf. Fig. 8-14 of Eccles, 1979a), and from various thalamic nuclei and cortical zones.

In Fig. 5-1 B the septal nuclei (S) are in the centre of the action. Of special importance is the medial forebrain bundle (MFB) with its two-way communication to the tegmentum and the hypothalamus. Also there is another descending pathway via the stria medullaris to the habenula nucleus and thence to the tegmentum and the interpeduncular nucleus. Thus the septal nuclei have good communication to the brain stem with the potentiality of evoking general bodily responses. In Fig. 5-1 A and B the hypothalamus is shown as background. However, it is vitally concerned in the expression of the emotions, being appropriately called the head ganglion of the autonomic system, which controls the cardiovascular system and the viscera. The hypothalamus also controls the secretions of the pituitary gland which act throughout the body as chemical messengers.

It has been erroneously believed that the human septal nuclei are atrophic, their position in the septum of lower mammals being occupied by a thin membrane devoid of nerve cells, the septum pellucidum, separating the lateral ventricles. However, by a rigorous comparative study Andy and Stephan (1968) have shown that there has been a ventral migration of the septal nuclei to form the 'septum verum'. The human septal complex is in fact very well developed having a regression index of 4.53, which is almost double that of simians and apes (2.19 to 2.40). The regression index expresses the size relative to that of a basal insectivore of an equivalent body weight.

Several other main pathways are shown in Figs. 5-1 and 5-2. The fornix (Fig. 5-1 A) is a great tract with more than a million fibres that is the main efferent pathway from the hippocampus and that circles around under the corpus callosum to end in the septal nuclei (S), the hypothalamus (H) and the mammillary bodies (M). In Fig. 5-1 B the mammillary body projects to the anterior thalamus, which powerfully projects to the cingulate gyrus (Fig. 5-2, AT to CG) and so to the neocortex. Fig. 5-2 also shows that pathways from the septal nuclei and the amygdala go to the large mediodorsal thalamic nucleus (MD) and so to the neocortex, mostly to the prefrontal lobe. In Fig. 5-2 there are several pathways from various

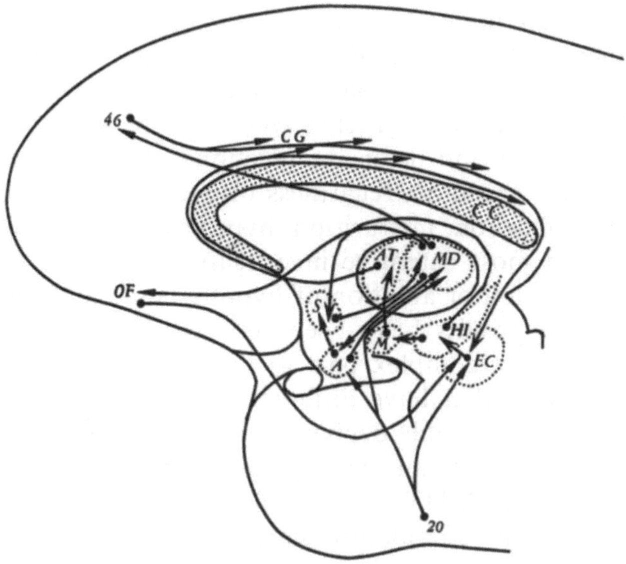

Fig. 5-2. Schematic drawing of the medial surface of the right cerebral hemisphere to show connectivities from the neocortex to and from the medio-dorsal thalamus (*MD*) and the limbic system. *OF*, orbital surface of prefrontal cortex; *HI*, hippocampus; *S*, septum; *CC*, corpus callosum; *EC*, entorhinal cortex; *A*, amygdala; *CG*, cingulate gyrus; *M*, mammillary body; *AT*, anterior thalamus; *46* and *20* are Brodmann areas, see Fig. 3-1

neocortical areas to the limbic system (cf. Fig. 3-2). These pathways presumably are of great significance in the reciprocal exchange of information between the neocortex and the limbic system, which is concerned in the development and expression of emotions.

This brief anatomical digression will be of value when attempting to account for the various reactions evoked both by stimulation and by ablation of sites in the limbic system and to relate them to the affects listed in Table 5-1. In this project the relatively few human experiments are of particular importance.

5.2.2 Limbic System and Emotional Expression

The pioneer work of Hess (1932) provided the first systematic study of the effects of stimulating in conscious animals the limbic and hypothalamic areas. The stimulating electrodes were implanted by a prior operation under anaesthesia. Great care was taken after the experimental study to localize the site of the stimulation. Hess showed that it was thus possible by stimulation to evoke emotional animal behaviour that was largely effected via the autonomic system. There could be fear and defensive reactions, snarling, hissing, piloerection, or there could be, regardless of need, eating and drinking, or vomiting, defecation, micturition, hypersexuality. Since that time there have been most extensive investigations into the effects produced by stimulation or ablation of hypothalamic and limbic areas. Often conflicting results have been reported. These arise because the area is the most complex part of the brain with diverse nuclei in close apposition that have different and often opposite actions. Also there is a considerable degree of species difference.

Fernandez de Molina and Hunsperger (1959) continued the work of Hess and Brügger (1943) on the defence reaction of the cat. They sharply localized the stimulation sites to the dorso-medial zone of the amygdala, the stria terminalis and the associated hypothalamus, the pathway being shown in Fig. 5-1 A. They state that

> the affective reaction most consistently obtained with low intensity stimulation was growling integrated in a defence pattern, the latter characterized by lowering of the head, flattening of the ears, dilatation of the pupils, and piloerection of back and tail. When stimulation was increased, the growls grew louder, or were followed by hissing, the defence pattern being accentuated by hunching of the back; sometimes the growling was followed by shrieking, or the growling-hissing response led to sudden flight.

It will be appreciated that these reactions are strikingly similar to the responses of a cat confronted by a dog, but they are not indicative of pain. In the cat the ventrolateral zone of the amygdala gave evidence of its dominant olfactory input with sniffing and searching as the most common reactions (Egger and Flynn, 1967).

Using the technique of chronically implanted electrodes Olds and Milner (1954) initiated a new approach to the study of emotions. It was found that with electrodes implanted so as to stimulate the septal nuclei or the medial forebrain bundle (cf. Fig. 5-1 B), and with arrangements so that the rats could stimulate themselves by pressing a bar, there ensued a behavioural extravaganza with up to 7000 self-stimulations an hour! Even the behaviourists had to recognize that they were confronted by a behavioural response which could best be described as due to stimulation of a pleasure centre.

Thus instead of the restrictive stimulus–response theories of behaviour Isaacson (1974) states that

> the experiences of the organism had to become a factor in describing or explaining behaviour. Pleasure and pain became accepted as legitimate terms, once again, for scientists trying to discover the neural basis of behaviour.

This changed attitude with respect to the interpretation of animal experiments was greatly encouraged by the reports of human experiments that will shortly be described. It was remarkable that the rats would forgo food or sex in order to perform this self stimulation, and would brave pain from foot stimulation in order to get to the stimulation device. Other sites in the limbic-hypothalamic region were less potent than the septal-medial longitudinal bundle zone. In yet other sites stimulation produced aversive responses that motivated the learning of an avoidance task. As already described for the cat, amygdala stimulation mostly gave aversive responses, but there was also a zone that when stimulated suppressed an attack behaviour.

Bilateral ablation of the septal nuclei usually resulted in the opposite of the pleasurable stimulation, namely irritability and aggression. With ablation of both amygdalae the animals tended to be tame and to exhibit hypersexuality often of a bizarre kind. Clemente and Chase (1973) give a detailed account of investigations of the limbic-hypothalamic structures concerned in animal aggression. It seems that the limbic system (amygdala and septum) acts by modulating hypothalamic actions for or against aggression respectively.

On rare occasions patients with disorders of the limbic system have provided the opportunity for investigating the effects of stimulation and ablation. It is satisfactory that the human experiments correspond fairly well with the animal experiments.

Delgado (1969) reports a remarkable case of amygdaloid disorder in a girl of 20 years who had suffered from encephalitis when 18-months-old:

> Her main social problem was the frequent and unpredictable occurrence of rage which on more than a dozen occasions resulted in an assault on another person such as inserting a knife into a stranger's myocardium or a pair of scissors into the pleural cavity of a nurse. The patient was committed to a ward for the criminally insane, and electrodes were implanted in her amygdala and hippocampus for exploration of possible neurological abnormalities ... demonstrated marked electrical abnormalities in both amygdala and hippocampus ... it was

> demonstrated that crises of assaultive behaviour similar to the patient's spontaneous bursts of anger could be elicited by radio stimulation of contact 3 in the right amygdala. A 1.2 milliampere excitation of this point was applied while she was playing the guitar and singing with enthusiasm and skill. At the seventh second of stimulation, she threw away the guitar and in a fit of rage launched an attack against the wall and then paced around the floor for several minutes after which she gradually quieted down and resumed her usual cheerful behaviour. This effect was repeated on two different days ... this finding was of great clinical significance in the orientation of subsequent treatment by local coagulation.

There have been other reports of a comparable kind. Mark and Ervin (1970) investigated the effect of amygdala stimulation on two patients with epileptiform seizures of the temporal lobe associated with uncontrollable violence. With a medial location of the electrode, stimulation evoked uncontrollable violence, but with a lateral location there were pleasant feelings, elated and floating and warm. Relief of the two patients was effected by surgical extirpation of the amygdala.

Heath (1954, 1963) reported that in a large number of schizophrenics stimulation through electrodes implanted in the septal nuclei gave agreeable feelings with sexual overtones so that some patients self-stimulated like Olds' rats! Delgado (1969) reports on three of his cases with psychomotor epilepsy. With electrodes implanted in the amygdala, presumably in the lateral zone, stimulation evoked pleasurable sensations with excessively friendly behaviour.

5.2.3 Conclusions on Limbic System

In summary we can adopt the simplified hypothesis of MacLean (1969, 1970) that there are two main components in the limbic system which correspond to the agreeable and disagreeable affects in Table 5-1. The septal nuclei, the medial forebrain bundle and the associated hypothalamus (cf. Fig. 5-1 B) are concerned with providing agreeable affects and the associated emotions that often have sexual overtones, as can be seen in the items listed under agreeable, general, in Table 5-1. The amygdala (medial zone) with its projections in part by the stria terminalis (Fig. 5-1 A) give the aversive feelings listed under disagreeable-general in Table 5-1. Excessive amygdaloid activity can have disastrous results on the patients, who may be classed as dangerous – even homicidal – criminals. I suspect that Gary Gilmour, the last criminal to be executed in the United States, may have been suffering from amygdaloid seizures. Unfortunately, there is so much prejudice against what is called psychosurgery that such cases go undiagnosed and untreated.

So far we have concentrated attention on the limbic system and the associated structures in a study of the generation of affects, but it must be recognized that affects (MacLean, 1970) are conscious experiences. In accord with the hypothesis of dualist-interactionism (Lecture 2) they are associated with the activation of modules in the neocortex in some unique spatio-temporal patterning. For example in Fig. 1-7 feelings or affects are shown under Inner Sense, and would be expressed as emotions by action on the appropriate modules. Thus the pathways between the limbic system and the neocortex provide an essential link in bringing about conscious experiences of affects and emotions. Figure 5-2 shows in outline some of the pathways. The medio-dorsal (MD) thalamic nucleus is a key structure since it receives from the amygdala and septum and projects very widely to the neocortex, in particular to almost the whole prefrontal lobe (Fig. 5-3). The anterior thalamus (Fig. 5-1 B) is also important by its projection to the cingulate gyrus (Figs. 5-2, 5-3) and from there widely to the neocortex.

5.2.4 Pharmacology of Limbic System and Hypothalamus

Serotonin and the catecholamines, noradrenaline and dopamine, are the transmitters most involved in influencing the limbic system and hypothalamus, and in thus modifying behaviour. Figure 6-7 shows the serotonergic pathways that project to the limbic system and the hippocampus. There is a similar projection for the noradrenaline pathways (cf. Fig. 8-14, M^c^Geer et al., 1978). There is good evidence that these amines play important roles in the mediation of emotional and behavioural states (Kety, 1970, 1972). Noradrenaline and dopamine seem to be specially concerned in arousal, aggression, self-stimulation and feeding, while serotonin gives relaxation and sleep as described in Lecture 6. In summary Kety (1972) states that

> it seems quite futile to attempt to account for a particular emotional state in terms of the activity of one or more biogenic amines. It seems more likely that these amines may function separately or in concert at crucial nodes of the complex neuronal networks which underlie emotional states. Although this interplay may represent some of the common features and primitive qualities of various affects,

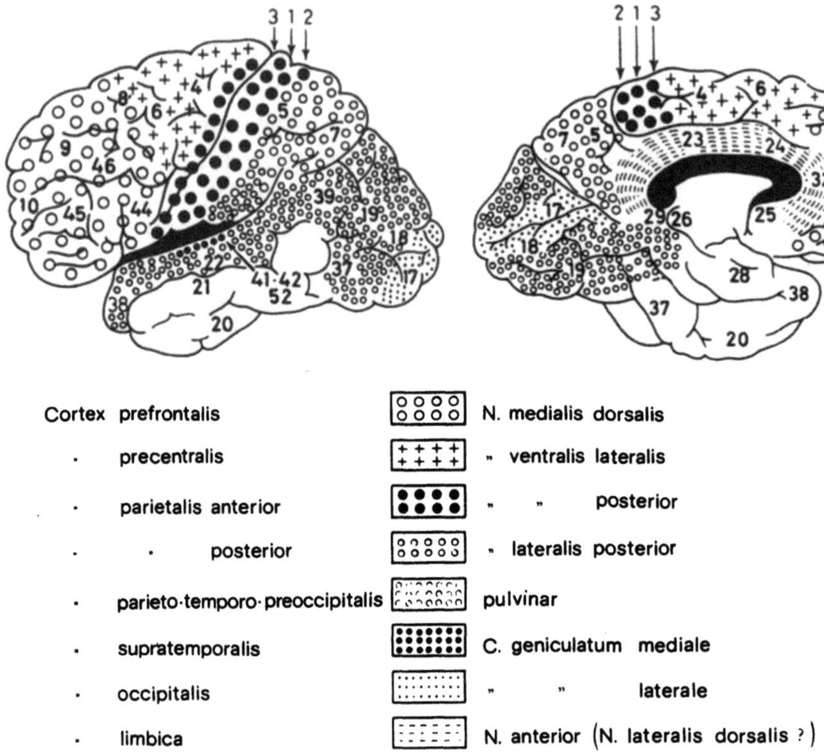

Fig. 5-3. Diagrammatic representation of the regional distribution in the neocortex of man of the thalamo-cortical connections. Left and right diagrams show respectively for the left hemisphere the lateral and medial aspects. The connections between the thalamus (*below right*) and the cortex (*below left*) are mapped in the hemisphere (*above*). It is to be noted that the MD thalamus is distributed to the whole prefrontal cortex while the anterior thalamus projects to the whole cingulate gyrus. Garey, L.J.: In: Handbook of Electroencephalogr. Clin. Neurophysiol. Vol. 2, 2A. Rémond, A. (ed.), p. 57. Amsterdam: Elsevier (1976)

the special characteristics of each of these states are probably derived from those extensions of the networks which represent apperceptive and cognitive factors based upon the experience of the individual.

The action of various pharmacological agents can be related to these transmitters. For example the hallucinogenic action of LSD seems to be related to serotonin receptor sites. The calming action of reserpine makes it an effective anti-psychotic, apparently by depletion of dopamine. For a detailed account of the pharmacology of the limbic system and the transmitters, reference should be made to Chapter 14 of M^C^Geer et al. (1978).

5.3 Pain

5.3.1 Introduction

Inspection of Table 5-1 reveals that of all the items listed there is one that is of outstanding importance in our enquiry concerning the emotional brain. Though all these affects can be experienced subjectively, pain is in a very special position. In the first place it is the most powerful generator of emotional states, as can be appreciated when one calls to mind acute agony or prolonged suffering. In the second place it is the only affect which has been subjected to systematic and concentrated study both in experimental animals and in suffering patients.

Our present interest must be concentrated on the problem of how activities in the neuronal machinery of the brain can generate this extraordinary range of disagreeable affects. Although affects are privately experienced, there is never any doubt about the reality of the pain except to the radical materialists and behaviourists. One is reminded of the limerick:

> There was a young lady from Deal
> Who said that, though pain isn't real,
> When I sit on a pin,
> And it punctures my skin,
> I dislike what I fancy I feel

By intersubjective communication the objective reality of pain is ensured during the whole range of our experiences from babyhood onwards. By this means pain can be investigated in conscious human subjects, and it is most important in clinical investigations both as a symptom and as a sign. Practical classes for medical students should include many kinds of investigation into pain. In my department in Dunedin, there were three 3-h practical classes on various kinds of pain inflicted on students by one another.

Before going into a scientific account of investigations on pain, it is desirable to consider the extraordinary nature of pain as a perception. All of the perceptions listed in Fig. 1-7 have some counterpart in the material world except pain – smell and taste with chemical substances, light and colour with electromagnetic waves and so on. It is not so with pain. There is no material counterpart as such. That extraordinarily disagreeable sensation is entirely made by us. We may ask – how did it have to be so disagreeable? How is it generated to be so nasty? But the same question of the nature of the perception can be asked about the other experiences listed in Fig. 1-7. For example how are certain wave lengths of electromagnetic radiation transmuted into colour and light, and how are frequencies of pressure waves converted into sounds? We come then to realize that the generation of pain in our brain–mind interaction is no more mysterious than the generation of the other types of perceptions in the frame OUTER SENSE in Fig. 1-7. Moreover there is the same basic similarity in that all are dependent on the stimulation of specific sense organs that cause the discharge of impulses that carry the coded information along nerve fibres going to the brain, where the perceptual experiences arise as a result of processes but dimly understood (cf. Lecture 3). The perceptual experiences are not in the brain, where there is the coded information, but in the conscious mind (World 2) as indicated in Fig. 1-7.

It will be realized that with consciously experienced pain a new phenomenon arose in the perceptual world. We may ask, biologically, why this should have occurred? Even in primitive nervous systems there are reactions ensuring withdrawal from injurious agents. There is no question of a felt pain giving this response. For example with a spinal frog the foot is rapidly retracted from an injurious acid solution by a spinal flexor reflex. We also carry out these automatic withdrawal responses from noxious stimuli. The hand or foot is rapidly withdrawn from a sharp or hot object before we feel the pain. Why should the unpleasantness of pain be superimposed on such an efficacious withdrawal response?

My suggestion is that a powerful signal was important in order to attract the attention of the subject so that a more efficacious and better organized response occurs and also so that memory of the damaging incident could lead to its avoidance in the future. I readily agree that this explanation is insufficient because many of the conditions giving great suffering were not amenable to treatment until the development of rational therapies. For example, as Popper has wittily remarked, what was the purpose of dental pain until the invention of dentists? And even today such severe pains as occur in the terminal stages of cancer can only be mitigated, not eliminated, and they serve no useful purpose in relation to the course of the disease. The best explanation we can offer is that pain is a signal to relax and rest the injured part, e.g. a damaged limb or a heart with failing circulation, and in this way to aid in recovery.

Another example of suffering is listed in Fig. 1-7 under INNER SENSE in the wide framework of 'feelings' (cf. the disagreeable column of Table 5-1). As we all know there are experiences of mental anguish which have no relation to injurious conditions in the material body. We may call them 'mental pain', as distinct from 'bodily pain'. Both these types of suffering have a deep underlying similarity, as is exemplified by the use of the words pain and suffering for each. We do not know to what extent animals participate in these vivid and traumatic conscious experiences. Mammals and birds certainly react to injurious stimuli as if they felt pain, and we must always strive to mitigate their presumed sufferings, just as we do for human beings. Mental suffering is really a mystery, a creation of the conscious mind without immediate reference to the happenings in the material world, but intimately related to memories. The experiential similarity between bodily induced pain and mental pain leads to the suggestion that related areas of the liaison brain are concerned (cf. Fig. 1-7).

5.3.2 Anatomy and Physiology of Nociceptive Pathways

This general philosophical introduction leads on to the scientific investigation of pain. Since the actual neural mechanism concerned in the generation of pain is not understood, the word nociceptive is used for pathways that are presumed to carry the information leading to the experience of pain. I will simplify the description to the essentials so that we can relate as well as possible the happenings in the brain to the experience of affect and the reaction of emotion. It will be recognized that the story as it can be told at present is very incomplete.

There are two distinct nociceptive systems in the skin and this distinction is preserved in the transmission up to the brain and to the perceptions derived therefrom, as is illustrated in Figs. 5-4 and 5-5. The distinction can be appreciated by a very simple test. A brief touch by the back of one's finger on a hot water pipe or other hot object results in an instant sharp pain followed by a slow and prolonged burning pain. The receptor organs stimulated by noxious agents are very primitive for both these kinds of pain, being

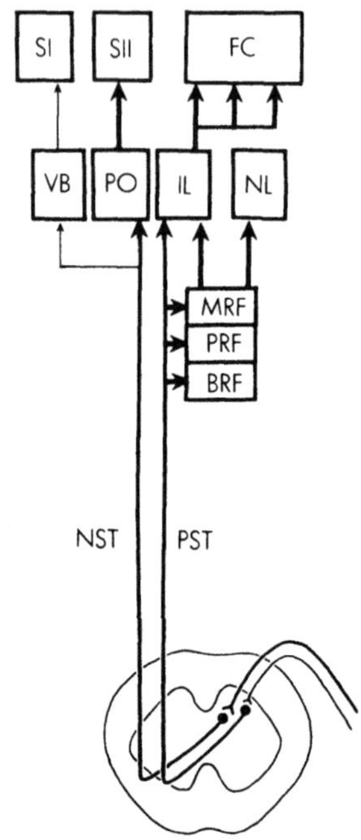

Fig. 5-4. Diagram to show the neospinothalamic (*NST*) and the paleospinothalamic (*PST*) pathways for nociception. The entry into the spinal cord is shown with a thicker myelinated fibre and a thinner non-myelinated fibre, but after relay in the spinal cord both pathways ascend in myelinated fibres. The various distributions in the brain stem and cortex are described in the text. *MRF, PRF* and *BRF* are respectively the mesencephalic, the pontine and the bulbar reticular formation. *VB*, ventrobasal thalamus; *PO*, posterior complex of thalamus; *IL*, interlaminar nucleus; *NL*, nucleus limitans; *SI* and *SII*, somatosensory areas I and II; *FC*, frontal cortex

apparently branching bare nerve terminals, though more refined examination will doubtless discover special anatomical features in these nerve terminals. In the system for the fast pricking pain small myelinated fibres carry the message to the spinal cord, which can be at a frequency of impulses up to 100/s for strong stimulation. In the system for the slow burning pain the message is in unmyelinated fibres (C fibres), which also can carry 100 impulses a second at the start of a severe stimulation (Iggo, 1959, 1974). Radiant heat provides a very convenient way of investigating the burning pain (Hardy and Stolwijk, 1966). However the slow pain also results from excitation of another quite distinct group of unmyelinated fibres with receptors stimulated by severe mechanical stress (Iggo, 1974).

In the spinal cord there are complex synaptic mechanisms for transmitting the nociceptive information, the excited nerve cells being in laminae I, IV and V of the dorsal horn (Iggo, 1974; Willis et al., 1974). In the human spinal cord there are two pathways up to the brain for nociceptive information, as is indicated in Figs. 5-4 and 5-5, both being on the contralateral side in the anterolateral column. It appears that most of the cells of origin of these tracts lie in the middle of the ipsilateral side, the intermediate nucleus, the axons immediately crossing to ascend in the two tracts (Réthelyi and Szentágothai, 1973). Both tracts are composed of small myelinated fibres, i.e., the unmyelinated peripheral input relays by a myelinated tract.

The neospinothalamic tract (NST in Fig. 5-4) carries the nociceptive information leading to sharp pricking pain, though only a fraction of that tract is so engaged. It terminates mainly in the posterior nuclear complex of the thalamus (PO) and to a small extent in the ventrobasal nuclear complex (VB), which is the relay station for somatosensory fibres in general, the DCN line in Fig. 5-7. After synaptic relay there is transmission, as indicated in Fig. 4-3, predominantly to the second somatosensory cortex, SII, which lies ventrolateral to the primary SI strip that is alone indicated in Fig. 1-1. Figure 5-6 shows the relation of the SII to the SI areas in a monkey brain, the fissure of Sylvius being opened out to show the full extent of SII (Jones and Powell, 1969). Figure 5-6 also shows the strong reciprocal communications between SI and SII, between both SI and SII and the primary motor cortex, area 4, and the weak one-way communication to the supplementary motor area, SMA. SI and not SII has a one-way communication to area 5 that is indicated in Fig. 3-2 (Jones and Powell, 1969, 1970b).

In the cat, Graybiel (1973, 1974) describes the projection of PO to the caudal part of SII and the adjacent cortex (Fig. 5-7) and also to the caudoputamen (Fig. 5-7) and the ventrolateral amygdala. Figure 5-7 can be correlated with Fig. 5-4 by recognizing that the stippled areas to the sides of VB are equivalent to PO in Fig. 5-4.

Poggio and Mountcastle (1960) studied in cats the responses of the PO and VB thalamic nuclear cells. In VB there were no nociceptive cells, whereas in PO 60% were nociceptive. This would suggest

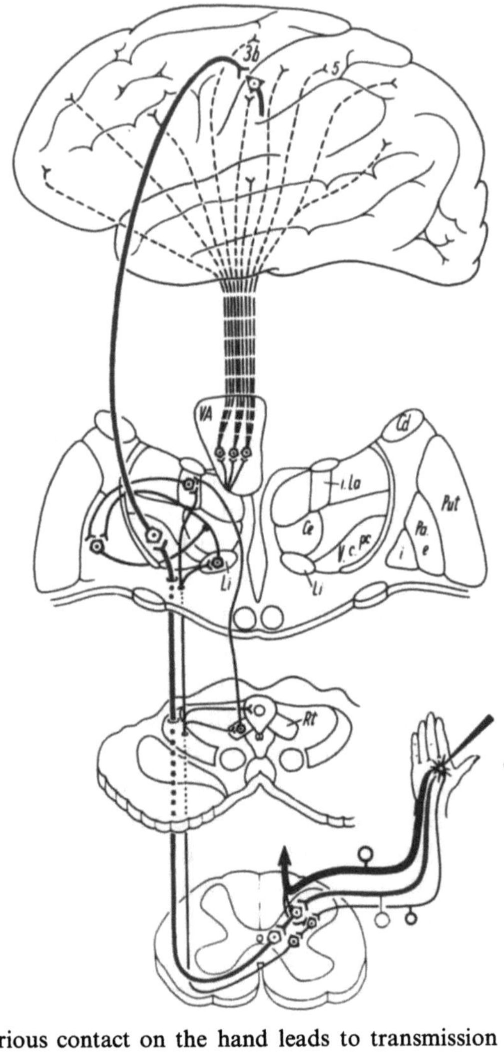

Fig. 5-5. An injurious contact on the hand leads to transmission in three pathways. One shown in *thick black* is for touch and it ascends in the dorsal column as shown by the *arrow*. The other two are the nociceptive pathways shown in Fig. 5-4, *grey* for neospinothalamic and *black* for paleospinothalamic. The branches of these two pathways in the brain stem are shown as described in the text: *Rt*, reticular formation; *Li*, nucleus limitans; *i.La*, intralaminar nucleus; *VA*, ventro-anterior thalamic nucleus; *Pa.e* and *i*, external and internal pallidum; *Ce*, nucleus centrum medianum; *Put*, putamen; *Cd*, caudate. *V.c.pc*, parvicellular ventro-caudal nucleus = *PO* of Fig. 5-4. Note pathway from Pa.e diving under Ce and i.La to project to VA and thence widely to the cerebral cortex. Hassler, R.: Central interactions of the systems of rapidly and slowly conducted pain. In: Advances in neurosurgery. Penzholz, H., Brock, M., Hamer, J., Klinger, M., Spoerri, O. (eds.), Vol. 3. pp. 143–150. Berlin, Heidelberg, New York; Springer 1975

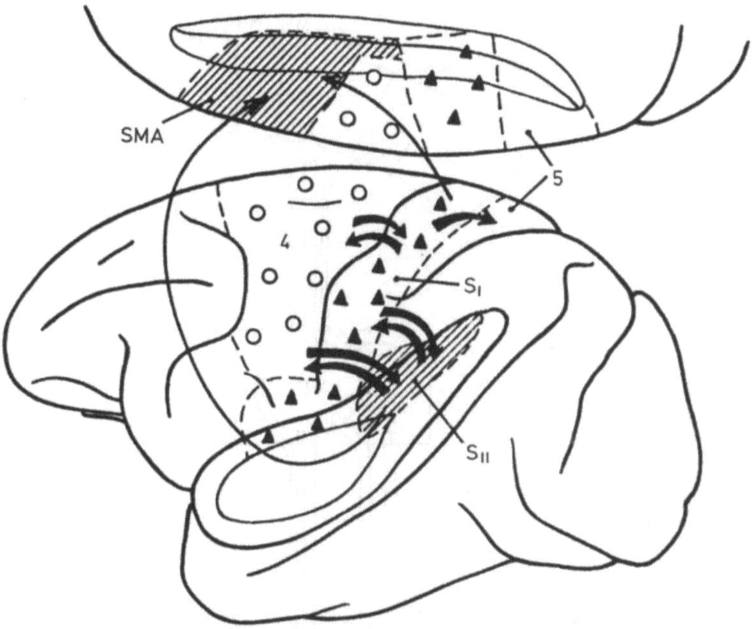

Fig. 5-6. A schematic diagram to show the anatomical pathways related to SI and SII in the monkey brain. Reciprocal connections join SI and SII to one another and to area 4, and small projections pass from SI and SII to the supplementary motor area (*SMA*), but only SI sends fibres to the parietal field, area 5. Jones, E.G., Powell, T.P.S.: Connections of the somatic sensory cortex of the rhesus monkey. I. Ipsilateral cortical connections. Brain *92*, 477–502 (1969)

that in Fig. 5-4 the input of the neospinothalamic path to VB contained no nociceptive fibres. Any nociceptive input to SI would be dependent on the small PO input. In fact it has often been maintained that SI has no nociceptive input and is not concerned at all with pain. However, Penfield and Boldrey (1937) reported pain from stimulating SI in 11 out of 426 patients, and Mountcastle and Powell (1959) in cats found 12 nociceptive in 593 cells. Similarly Carreras and Andersson (1963) found 5/262 nociceptive cells in SI and 16/268 in SII in cats. These various findings must be considered in relation to the important discovery by Whitsel et al. (1969) that SII in monkeys is divided into two distinct regions, there being a small posterior strip that was particularly responsive to noxious and auditory stimuli. This would correspond to a dominant activation of this strip by the PO thalamic nuclei, where there is likewise a selective responsive-

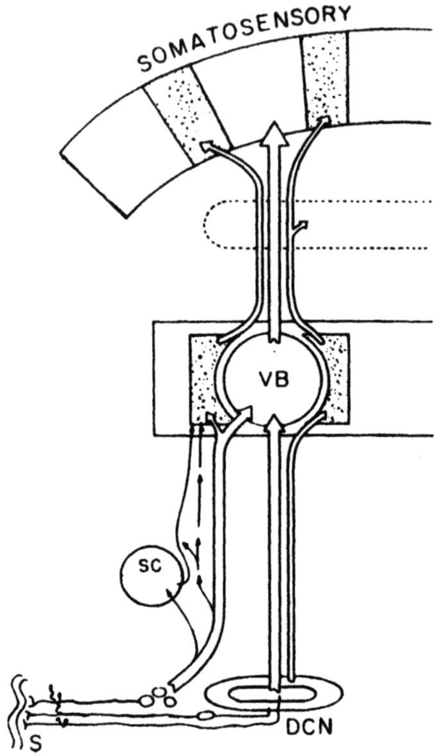

Fig. 5-7. Highly schematic diagram illustrating ascending somatosensory pathways. At the bottom are sensory elements from the skin (*S*) that ascend via the dorsal column nuclei (*DCN*) to the ventrobasal thalamus (*VB*) and thence to the somatosensory cortex by the *large arrows*. To the left are the spinothalamic pathways that ascend to the superior colliculus (*Sc*) and to thalamic areas (cf. Figs. 5-4 and 5-5) shown *stippled* on each side of VB and thence project to areas of the somatosensory cortex (shown *stippled*) other than the areas 3, 1, 2 for the VB projection. Note also projection to the area outlined by a *broken line*, which represents the basal ganglia (cf. Pa.e of Fig. 5-5). Graybiel, A.M.: Studies on the anatomical organization of posterior association cortex. In: The neurosciences. Third Study Program. Schmitt, F.O., Worden, F.G. (eds.), pp. 205–214. Cambridge, Mass.: MIT Press 1974

ness to noxious and auditory stimuli. A related finding in the monkey is that the medial part of the posterior complex of the thalamus (PO) projects to the retroinsular cortex lying posterior to SII (Burton and Jones, 1976).

There are reports that ablation of SII in cats and man results in a diminution of pain sensitivity and that relief of severe pain in patients occurs (cf. Mountcastle, 1968). Most surgical ablations were confined to the easily accessible SI, and, as might be expected, were disappointing. Stimulation of the posterior nuclei of the thalamic complex, nucleus limitans in particular (cf. Li in Fig. 5-5), produced severe pain (Hassler, 1960).

The paleospinothalamic tract, PST (cf. Figs. 5-4 and 5-5), carries the slow-burning nociceptive pathway and is quite separate, on the medial side of the neospinothalamic tract. It gives off collaterals to the reticular formation in the medulla, pons and mesencephalon, BRF, PRF, MRF, in Fig. 5-4 (Bowsher, 1965; Albe-Fessard and Besson, 1973) before terminating in the nucleus limitans (NL, Fig. 5-4; Li, Fig. 5-5) and in the intralaminar nuclei (IL, Fig. 5-4; iLa, Fig. 5-5). The reticular relay also goes to the intralaminar nucleus and the nucleus limitans after sending collaterals to the hypothalamus. From the intralaminar thalamus there is a widely dispersed thalamic projection to the cerebral cortex (cf. Jones and Leavitt, 1973; Graybiel, 1973). The mesencephalic reticular formation plays a major role in the transmission of nociceptive impulses in cats (Melzack et al., 1958).

In Fig. 5-5 Hassler (1975) illustrates an important pathway from the paleospinothalamic tract to the neocortex. The thalamic relay is via the nucleus limitans (Li) and the intralaminar nucleus (iLa), both of which project to the external globus pallidus (Pa.e) of the corpus striatum, which is shown projecting to the VA thalamus and thence being widely dispersed to the neocortex. If the neospinothalamic pathway to the cortex is interrupted, the paleospinothalamic pain becomes much more severe and even arises spontaneously. This so-called protopathic pain can be treated surgically by coagulation of the nucleus limitans (Hassler, 1975).

There has tended to be a suspicion that the neocortex is not concerned in the perception of pain. All the results from noxious stimulation could be interpreted as being due to other skin receptors stimulated at the same time. It is therefore of special interest that the stimulation of purely pain receptors in the tooth pulp evokeds cortical responses in SII with a latency much longer than for infraorbital nerve stimulation (Vyklický et al., 1972; Andersson et al., 1973).

Brief reference should be made to a third pain system, namely that generated by nociceptive pathways from viscera and muscle, including the heart. This pain corresponds closely to the slow-burning skin pain and the nociceptive pathway is up the same paleospinothalamic tract. Often there is convergence onto the same projecting cells with the consequence that this deep pain is often referred to some cutaneous zone in conformity with the convergent lines.

At all synaptic relays in the nociceptive pathways there is opportunity for inhibitory suppression of the transmitted information. At the level of the synaptic relays in the spinal cord (Fig. 5-4) presynaptic inhibition is dominant, but postsynaptic also is effective (Schmidt, 1973). Also in the thalamic relay, postsynaptic inhibitory control of transmission is very effective. It is important that these inhibitions of the nociceptive pathways can be exercised by the cerebral cortex by means of cortico-thalamic and cortico-spinal pathways (Schmidt, 1973). Wall (1978) utilizes the various inhibitory controls on the nociceptive pathways at the spinal level in his very tentatively revived theory of gating control.

There is much interest in the inhibitory control of nociceptive pathways. Several clinical treatments activate this inhibitory mechanism in order to alleviate pain. It also provides a scientific explanation of the anaesthesias of acupuncture, hypnosis and counter-irritation. It is a familiar experience that when intensely engaged, even severe injuries are disregarded, presumably because of the inhibitory blockade by descending inhibitory controls. However, an alternative explanation is that selective attention is concentrated elsewhere as described in Lecture 4. It would be an important evolutionary design to have protection of the cerebral cortex from distracting inputs in survival situations.

5.3.3 Pharmacological Control of Pain

There is an immense amount of recent literature on the pharmacological control of pain that has been reviewed by Kerr and Wilson (1978) and by Fields and Basbaum (1978). It is too early yet to distil a clear and comprehensive story from the experimental evidence which is so various. However, there is general agreement on one most significant advance.

It has been established that morphine and related analgesic drugs exert their influence on the membranes of nerve cells at specific receptor sites. These opiate receptors, as they are called by Pert and Snyder (1973), occur in regions related to the nociceptor pathways, e.g. the amygdala, the hypothalamus, midbrain structures such as the periaqueductal grey, the habenula and the interpeduncular nucleus (cf. Fig. 5-1 B) and the dorsal horn of the spinal cord (Fig. 5-5). It was anticipated that such receptor sites should belong to a transmitter system acting in an analgesic manner. Very soon two such transmitters were identified (Hughes, 1975) as two small polypeptide molecules of only five amino acids. The locations of these enkephalins, as they are called, is in good accord with their proposed physiological action as transmitters onto the opiate receptor sites in the limbic system, the hypothalamus, and the midbrain structures. It was puzzling to find the largest concentrations in the globus pallidus, and the caudate nucleus. However, the globus pallidus concentration is in accord with Hassler's pathway shown in Fig. 5-5. The enkephalins act to depress neuronal activity, and it can be anticipated that they are transmitters of an inhibitory system that acts as a natural control on nociceptive pathways. They offer great opportunities for a therapeutic control of pain. The cells of origin are not yet identified.

5.3.4 Neural Substrate of Pain?

This question is of the greatest importance, yet at the best we can only make speculative answers. There is a temptation to give a location in the cerebral cortex, particularly when somatotopic localization is given as accurately as it is for the sharp pricking pain, for example by a needle prick or an insect bite. However, it is possible to argue that the localization is given by the associated mechanical stimulation. There is even some ability to localize the burning pain, as is given for example by a beam of radiant heat. However the skin pain associated with inflammation is always better localized when tested by probing.

In the perceptual experience of touch it has been shown (Libet, 1973; Libet et al., 1979) that a considerable development of the responses in the brain, for up to 0.5 s, is necessary for the most elemental sensory experience. As already mentioned in Lecture 4 sensory experiences do not arise until there is the development of complex patterns of neuronal action. The pathways for looping transmission between areas of association cortex have been investigated anatomically by Jones and Powell (1970b) in particular and by Mountcastle and associates physiologically (Mountcastle et al., 1975).

With the nociceptive input from the neospinothalamic tract largely to SII (Fig. 5-4) there is little opportunity for this wide cortical dispersal (cf. Fig. 5-6) except by way of SI, which seems improbable because SI is so deficient in nociceptive neurones. For that reason the pathways shown in Fig. 5-7 are important, for there is transmission to cortical areas adjacent to SII and also to the basal ganglia. In addition Graybiel (1973) has made the important discovery that the thalamic PO group projects to the amygdala. We have here a nociceptive path activating a limbic affect system of adversive nature. Moreover Graybiel (1973) suggests that the cortical regions adjacent to SII in the cat can be analogous to the superior temporal sulcus (STS) of primates, which has a wide projection to the prefrontal lobe and also to the limbic system (Fig. 3-2, STS).

It is most important to discover nociceptive pathways to the prefrontal cortex because that cortex is certainly of major significance in generating the 'affect' of pain perception. It is of particular interest that the paleospinothalamic tract is shown in Fig. 5-5 to project widely to the neocortex after relays in the thalamus (Li and iLa) then in the globus pallidus and finally in the VA thalamus.

In the very numerous and regrettable clinical treatments by prefrontal leucotomy there has been a remarkable separation of pain perception from affect. The patients still feel pain, but with little or no experience of affect. Simply expressed, they recognize the pain, but it does not bother them by hurting. Prefrontal leucotomy similarly relieves mental pain. It can be suggested that the prefrontal component in both physical and mental pain gives these pains their experiential similarity. To some extent also relief from the affect of pain occurs in right parietal lobe lesions, where the patients take little notice of the left side of their bodies, in what Hecaen (1967) appropriately calls 'a pantomime of massive neglect'.

Let us now enquire into the possible interrelationship of the limbic system and the neocortex in the generation of a painful experience from a nociceptive input. The projections of the medio-dorsal (MD) and anterior thalamic nuclei to the cerebral cortex indicated in Figs. 5-2 and 5-3 provide a powerful and widespread cortical input from the limbic system. Similarly there are powerful inputs from the cerebral cortex to the limbic system that are partly indicated in Figs. 3-2 and 5-2. Thus there is opportunity for reverberatory circuit operation between the neocortex and the limbic system.

In fact the tremendous disturbance of cortical function in the affect of severe pain makes it seem likely that the cerebral cortex is being occupied in intensely running reverberatory circuit operations that are built up by adjuvant inputs from the limbic system. The widespread cerebral activity in even moderate pain is well indicated by the studies on cerebral circulation (Lassen et al., 1978) that will be considered in the next lecture. The increased circulation in the frontal and parietal areas by moderate pain is seen in Fig. 6-2. An intense cerebral excitation would also be brought about by the reticular activating system. The paleospinothalamic pathway activates this system directly (Fig. 5-4) and also the hypothalamus. Moreover from the intralaminar and limitans nuclei of the thalamus there would be, as already described, a diffuse distribution of nociceptive information to the globus pallidus externa from which there are pathways to the VA and VL thalamic nuclei (Fig. 5-5) and so to the cortex (Hassler, 1978).

What then are the special attributes of the cerebral activity that give the mental experience of pain by interaction across the frontier of Fig. 1-7? I would suggest that three factors are concerned; firstly, a widely dispersed reverberatory activity of modules of the prefrontal and parietal cortices; secondly, reverberatory circuits involving on the one hand aversive limbic structures, such as the amygdala and associated hypothalamic areas, and on the other the cerebral cortex (Figs. 3-2, 5-2); thirdly, strong reinforcement from the reticular activating system (cf. Fig. 5-4). These suggestions are in line with the general belief that pain arises from most complex cerebral operations, which greatly complicates the difficulties attending the surgical relief of pain. It is suggested that on dualist-interactionism there is unification of the pain experience from the immense diversity of the neuronal operations in a large part of the brain. These neuronal operations must have a special property of urgency, perhaps because of the intensity of the modular operations (Fig. 1-7) activated by reverberatory circuits from the limbic system.

5.4 General Consideration of Other Affects

Finally we come to the other affects of Table 5-1 and the associated emotions. Hunger and thirst have been much less investigated than pain and they raise no special story that is of interest for our general purpose of trying to understand the emotional brain. Just as with pain we are far from being able to understand how the observed cerebral activity gives the experienced affect. This failure is particularly evident with the general affects. It is here that our knowledge of the emotional brain is so inadequate.

For example, I will take an extreme case that is known to us all. What can we say about the happenings in our brain when we are in love? We would I hope agree that we would distinguish love from sex in its anatomical, physiological and psychological manifestations, though there is a lamentable tendency at identification in these unchivalrous times. I am not denying the sexual aspect of love, but merely saying that love is much more than sex and is not only sex. The best illustrations of this distinction can be provided by the great literary tradition of love as displayed in poems, in plays and in novels. Love is a deep relationship between two human beings replete with remembrances, dedication, sacrifices and ideals. So often in literature love is unfulfilled or lost after an ecstatic fulfillment. Then it may become a source of longing and even despair. But we may be fortunate in being able to relive and enjoy our lost love in memory and imagination. Denis de Rougement (1972) has written with deep insight and scholarship on the theme of love in his great book *Love in the Western World.* He develops the theme that the story of Tristan and Isolde has been the model of so much love literature through the centuries. But I now intrude on the subject matter of Lecture 8, where love will be considered as a very special aspect of altruism. In true love consideration goes to the loved one and not to oneself, so it is altruism of a high order.

Lecture 6

Brain Activity and Levels of Consciousness

Résumé

There is a general introduction followed by an account of the principal methods of assessing human brain activity.

A most important technique uses *radio-Xenon* in quantifying and mapping the cerebral circulation. The findings of this method have been confirmed by *radio-oxygen* studies on patients. There is general agreement between these recent observations and older findings on metabolism and blood flow for the whole brain.

Electroencephalography also provides valuable data on the levels of cerebral activity.

There is firstly the classical work on the *reticular activating system* that via a relay in the non-specific thalamus induces cerebral activation as indicated by desynchronized EEGs. Two other pathways to the cerebral cortex are described because of their postulated implication in sleep: *serotonergic* that depresses cortical cells and *noradrenergic* that tends to cause arousal.

The most important and ubiquitous cause of unconsciousness is *sleep*, yet despite intensive studies the relationship of cerebral activity to sleep is enigmatic. It has long been known that oxygen consumption and blood flow are not reduced in sleep and may even be increased in the phases of deep sleep called *paradoxical sleep*. There is an account of the EEGs at various levels of sleep and of the spontaneous activity of cortical neurones during stages of sleep.

There is good evidence that sleep is induced and maintained by the serotonergic system, but it seems likely that the noradrenergic system is concerned in the phases of deeper sleep with the associated high levels of neuronal activity. But complex neuronal systems have to be envisaged to account for all the phenomena of sleep regulation on a diurnal cycle. There is brief reference to the pathological disorders: narcolepsy and hypersomnia. The relation of dreams to the rapid eye movements (REM) and the associated EEG is described and hallucinations are briefly considered.

If sufficiently generalized, convulsions result in loss of consciousness. The opposite state of greatly depressed neuronal activity also results in the unconsciousness of coma or of surgical anaesthesia.

On the hypothesis of dualist-interactionism, unconsciousness can plausibly be explained by the low level of cerebral activity in coma and anaesthesia and by the high driven level in convulsions. In these situations the liaison between the self-conscious mind and the spatio-temporal patterns of modular activity would be deteriorated or would fail altogether, hence the unconsciousness. However, the unconsciousness of sleep cannot so simply be explained. Possibly it is the changed temporal pattern of neuronal activity that is responsible. Dreams occur when there are changes in that pattern. The necessity for sleep remains an enigma. Moruzzi suggests it is a time for plastic reconstruction of the cortex after the intense activity of the waking day.

6.1 Introduction

Levels of consciousness provide a most important criterion in our enquiry into the relationship between cortical activity and consciousness. The question is in general terms: how far can neuronal activity in the brain be correlated with levels of conscious awareness? This enquiry can be pursued with respect to a wide variety of cerebral activities and the whole range of conscious awareness that may be assessed clinically.

Several criteria can be used for assessing the level of neuronal activity, and there is on the whole reasonable agreement between these criteria. However, it has to be recognized that our methods are still very inadequate, particularly in respect of the detailed patterned activity of cortical neurones. At the best we can sample the impulse discharges of a few neurones by simultaneous recording, which gives very inadequate data on the patterns of activity; and these presumably are the crucial factors. In any case the total number of cerebral neurones adequately investigated is not more than several thousand in mammalian brains. It is certainly *not* the case that the higher the level of neuronal activity, the higher the states of consciousness. It seems expedient to begin with the rather simple criteria of brain activity provided by studies of its metabolic level and the related cerebral blood flow before we become involved in the more complex phenomena of electrophysiological events.

6.2 Levels of Metabolic Activity in the Brain

Our enquiry is not concerned with the immense complexities of the biochemical processes in the brain. For our present purposes it is enough if we study the outcome of these metabolic activities in the cerebral levels of activation which can be measured by the oxygen consumption and the resulting changes in the blood circulation rate. The former measurements have been made for the whole human brain by ingenious gas inhalation techniques (Kety, 1961). However Lassen et al. (1978) report on a new procedure in which radioactive isotopes of oxygen were injected into the brain arteries by Raichle (1975) so that regional alterations in uptake of oxygen could be measured in patients. Another new technique for local metabolic study has utilized radioglucose injections (Reivich et al., 1975), but it can only be applied to animals because it depends on autoradiography of brain slices.

The pioneer investigation on oxygen consumption of the human brain utilized a nitrous oxide inhalation method. The mean value of 5.4 ml O_2/100 g brain/min would be in good agreement with the mean value now given by Ingvar (1975a) for the blood flow, 52 ml/100 g/min. Table 6-1 shows that under pathological conditions the oxygen consumption is reduced, even down to one-half. Ingvar (1975b) reports blood flows through deteriorated brains as low as 35 ml/100 g/min. Contrariwise the cerebral blood flow can rise by 20% with emotions, mental activity or with stimulation of afferent inputs into the brain (Ingvar, 1975a). Kety (1961) reported a 20% increase in cerebral blood flow during sleep (cf. below), but the cerebral oxygen consumption was not appreciably altered. Increases of cerebral blood flow as high as 50% were found in paradoxical sleep (see Seylaz et al., 1975 for details). The most spectacular increases occur in epileptic seizures, where in rats there may be a two- to

Table 6-1. Overall oxygen consumption of the brain in conditions of impaired consciousness[a]

	% of normal values		% of normal values
Alert, conscious	100	Surgical anaesthesia	64
Senile psychosis	82	Insulin coma	58
Diabetic acidosis	82	Diabetic coma	52
Insulin hypoglycaemia	79	Alcoholic coma	49
Artificial hypothermia	67		

[a] Kety, S.S.: Sleep and the energy metabolism of the brain. In: The Nature of Sleep. Ciba Foundation Symposium. Wolstenholme, G.E.W., O'Connor, M. (eds.), pp. 375–385. London: Churchill 1961

fivefold increase in brain metabolism and an increase of up to sevenfold in blood flow (Plum and Duffy, 1975). Ingvar (1975a, in *Brain Work*, p. 230) reported a radio-Xenon investigation on a patient where the cerebral blood flow rate of 40 to 50 ml/100 g/min was increased to 100 ml/100 g/min for the area involved in the seizure.

These measurements for the whole brain show that there is on the whole a good correlation between the levels of brain metabolism as measured by the oxygen consumption and the brain circulation. It is therefore of great interest that Ingvar, Lassen and others (Ingvar, 1975a, 1975b; Lassen et al., 1978) have perfected a radiotracer technique in which radio-Xenon is injected into the cerebral circulation. In this way, by a large battery of radiation detectors assembled over the scalp, the circulation can be simultaneously measured over as many as 254 areas of the human cerebral cortex under a wide variety of conditions.

Figures 6-1, 6-2 and 6-3 are derived from these 1 cm^2 plots of the circulation rate in the original grid. The local circulation rates are converted on a size-coded scale as shown below Fig. 6-1, the increases and decreases being indicated by solid and open circles, respectively. The resting pattern gives increases or decreases above the overall mean value, while in each other figure of Figs. 6-1, 6-2 and 6-3, the sizes are shown relative to the mean resting level, i.e., relative to the same base line as for the resting plots. At rest the circulation to the frontal areas is seen to be above the average levels for the brain (filled circles), while in the temporal and parietal it is below (open circles).

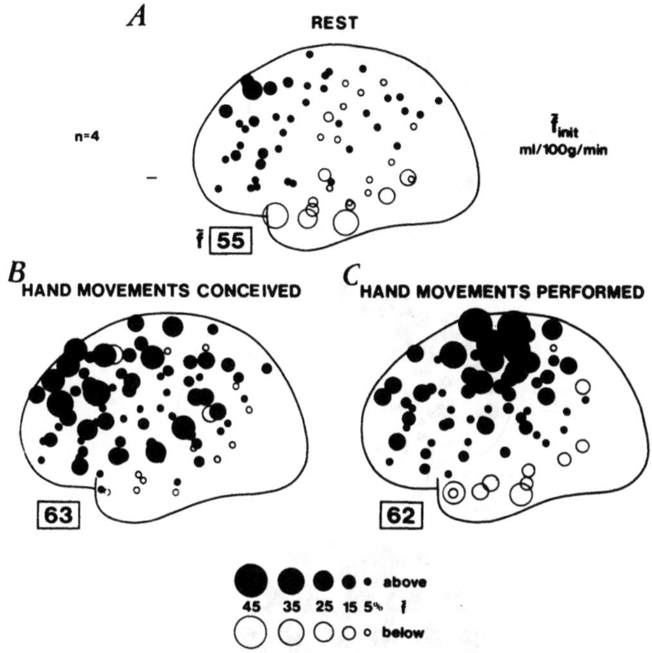

Fig. 6-1 A–C. Effects of intention and of performance of hand movements upon the regional cerebral blood flow (rCBF). The rCBF was determined by the radio-Xenon technique as described in the text. The results from four patients are plotted in a drawing of the left hemisphere. At rest some regions have a blood flow greater than the mean (*solid circles* in frontal) and some below (*open circles* in temporal). Percentages above and below are plotted on the size scale shown below. The resting blood flow to the brain is 55 ml/100 g/min, and it increases to 63 and 62 in the other two drawings. Ingvar, D.H., personal communication, 1979

Figure 6-1 B is most interesting in that it shows a considerable frontal increase when thinking of hand movements without moving the hand. When the hand movements were actually carried out, there was a large increase over the hand motor area (Fig. 6-1 C). In Fig. 6-2 B with a mild thumb stimulus there was an increase above resting level in the precentral flows close to the thumb area (cf. Fig. 1-1). In C a strong stimulus giving a mild pain evoked a large increase in blood flow over the whole frontal cortex and also into the anterior parietal zones, but it was still low in the temporal and posterior parietal areas. The overall cerebral circulation was now much increased to 55 ml/100 g/min from the resting level of 49 ml. Evidently a nociceptive input produces a large widespread circulatory increase. It can be assumed that this is indicative of an increased

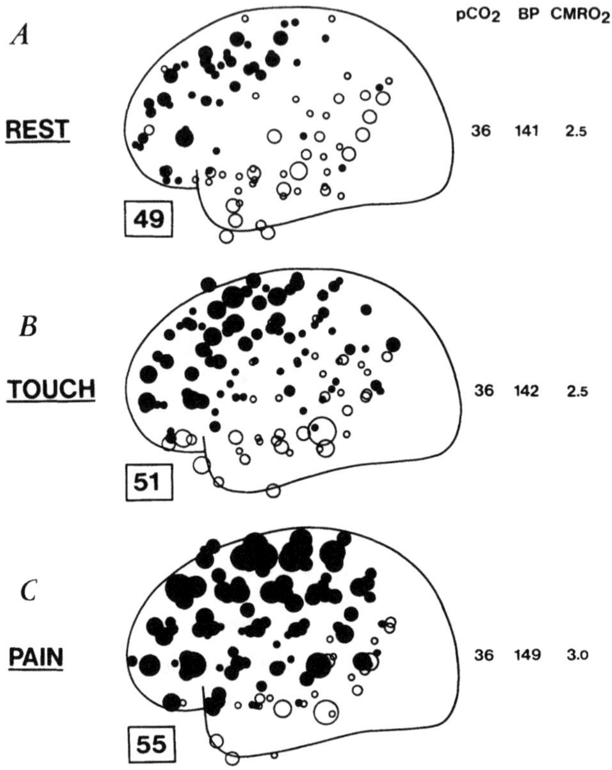

Fig. 6-2 A–C. Effects of touch and pain on the rCBF. The touch was applied to the right thumb at twice threshold intensity. The pain was caused by a high intensity stimulation evoking discomfort and pain. Note increases in cerebral blood flow from 49 to 51 to 55 ml/100 g/min. To the right are measurements of carbon dioxide pressure (pCO_2) of alveolar air, blood pressure (*BP*) and of the regional cerebral metabolic rate for oxygen ($CMRO_2$) in ml/100 g/min. Ingvar, D.H., personal communication, 1979

cortical metabolism associated with intense neuronal activity, as was proposed in Lecture 5.

Figure 6-3 is a simplified diagram summarizing several types of experiments (Ingvar, 1975a). There are plotted only the regions where the blood flows were 20% above (filled circles) or 20% below (open circles) of the mean brain values. Shading is employed to pick out the main zones of influence. As in Figs. 6-1 A and 6-2 A the resting brain has an excess of prefrontal circulation relative to all other areas. The figures labelled sens 1 and sens 2 are for touch and mild pain respectively, there being an increased precentral circulation with

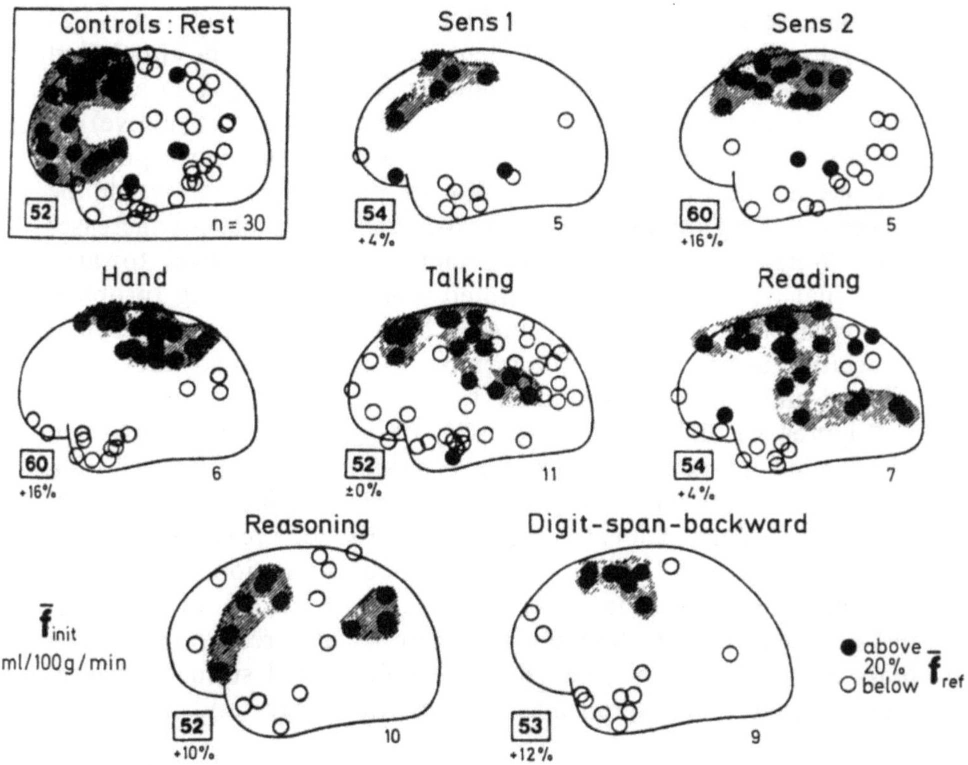

Fig. 6-3. Patterns of brain activity. Summarizing diagram of seven modes showing only deviations above 20% and below 20% of the hemisphere mean. The values in boxes denote the blood flow ($\bar{f}$) to the brain in ml/100 g/min. Percentages below boxes give mean increases above resting. Major changes in regions have been indicated by shading and are described below: at rest the rCBF distribution is distinctly 'hyper-frontal'; during low and high intensity contralateral cutaneous stimulation (sens 1, sens 2) there is a precentral flow activation which increases with the stimulation intensity; during contralateral voluntary hand work the main flow increase takes place over the rolandic and parietal regions; during speech and reading a Z-like activation pattern is induced over premotor, rolandic, and the sylvian region. During reading the lower part of the Z is especially marked; problem solving which includes visual activity (reasoning) augments the flow over pre- and postcentral association cortex; if visual activity is not involved in problem solving (digit-span-backward test) only a premotor activation is seen. Ingvar, D.H.: Patterns of brain activity revealed by measurements of regional cerebral blood flow. In: Brain work. Ingvar, D.H., Lassen, N.A., (eds.), pp. 397–413. Copenhagen: Munksgaard 1975

prefrontal spread (cf. Fig. 6-2 B and C). The hand figure (cf. Fig. 6-1 C) shows that voluntary hand movement greatly increases the circulation to the hand area of the sensorimotor cortex (cf. Fig. 1-1) and the adjacent precentral and parietal areas. Raichle et al. (cf. above) by a radio-oxygen technique have shown that there is a corresponding increase in the oxygen consumption of these areas. As would be expected talking and reading in Fig. 6-3 produce circulatory increases in the Wernicke area (cf. Fig. 1-1), and, with reading, back towards the visual occipital areas. It is interesting that reasoning (problem solving) increased the blood flow in both the parietal and frontal association areas, whereas the memory task (Digit-Span-Backward) affected only the precentral area.

Figures 6-1, 6-2 and 6-3 exemplify the extraordinary manner in which neuronal activities give rise to specific regional blood flows. This information will be of great value when considering the neural events that relate to mental events. It is also of great interest that in the resting state there is a higher blood circulation through the prefrontal areas than in other parts of the brain. A disability of this surface recording technique is that important areas of the human brain are not accessible, namely the whole medial surface (cf. Fig. 3-1), as well as the prefrontal orbital, the insula and the whole limbic system.

It is generally assumed that the activity of a neuronal system is principally reflected in the impulse discharges of the constituent neurones. Creutzfeldt (1975) has made an interesting calculation of the total heat production and therefore of the oxygen consumption of an area of the human cortex that would be attributable to impulse discharges of the constituent neurones. He has assumed reasonable values for the neuronal population and neuronal size, for the average impulse discharge rate (10/s) and for the heat production per impulse. The calculation shows that no more than 2% to 3% of the heat production and oxygen utilization (5.4 ml/100 g/min) can thus be accounted for. This is not surprising because it would be expected that the greater part of the energy utilization would be involved in operating all the metabolic machinery of the nerve cell. And in any case, this calculated energy figure is only for the initial energy generation by the impulse and not for the recovery process. Even the synaptic mechanisms are not taken into account. Nevertheless these calculations are of interest in showing how impulse generation in nerve cells is related to the oxygen consumption.

6.3 Electroencephalographic Studies of Neuronal Systems that Modify Cerebral Activity

6.3.1 Introduction

In Fig. 6-4 there are specimen electroencephalograms in various conditions associated with unconsciousness. They should be compared with Fig. 4-1, which shows the alpha rhythm of the relaxed state and the desynchronized responses occurring with various disturbances giving a higher level of cerebral activity. There will be further reference to these conditions, but our present concern is that there is a fairly good agreement between the electrical wave forms (cf. Fig. 6-4) and the cerebral metabolism. As indicated in Table 6-1,

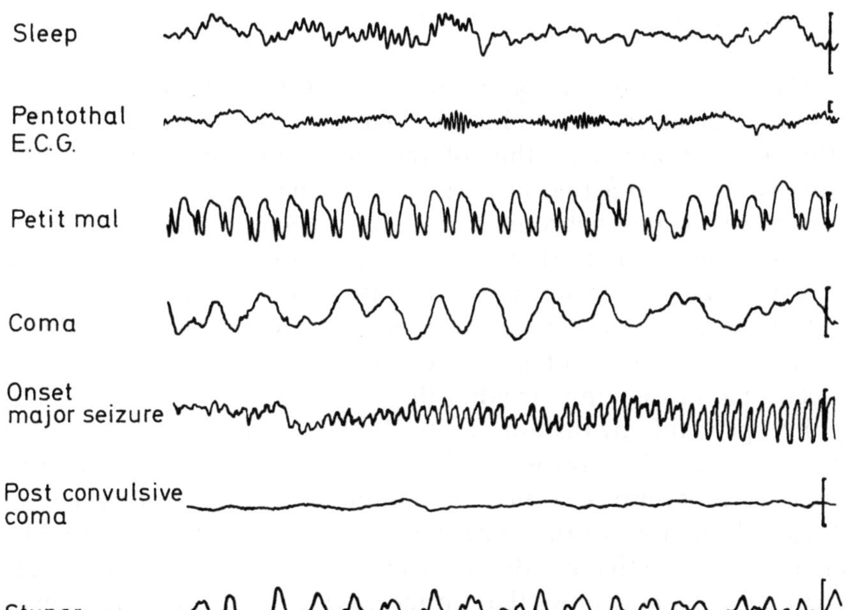

Fig. 6-4. Sample human electroencephalographic patterns taken during states of unconsciousness due to different causes. In all instances, all parts of the cerebral cortex of both hemispheres are involved, with regional differences in the form of abnormality in some instances. Jasper, H.H.: Pathophysiological studies of brain mechanisms in different states of consciousness. In: Brain and conscious experience. Eccles, J.C. (ed.), pp. 256–282. Berlin, Heidelberg, New York: Springer 1966

low oxygen consumption occurs in conditions with slow wave forms in Fig. 6-4 (coma and post-convulsion coma) and also with the great reduction in neural discharge in anaesthesia (Fig. 6-15). On the other hand, the strong driven rhythms of epileptic seizures in petit mal (Figs. 6-4 and 6-12) and of major seizures (Fig. 6-4, row 5) have strong neuronal activation (Fig. 6-14) and a correspondingly high oxygen consumption and brain circulation rate, even at several times the normal level.

The relation of impulses to levels of consciousness and unconsciousness will be dealt with at the end of this lecture. Before considering sleep, the most important state of unconsciousness, it is desirable to give an account of the mechanisms responsible for regulating the levels of awake and alert states.

6.3.2 Reticular Activating System

In 1949 Moruzzi and Magoun found that by stimulation of the reticular nucleus in the midbrain (in the region marked by TEG in Fig. 6-5) the alpha rhythm of the electroencephalogram (EEG) was converted into a fine fast wave form characteristic of arousal, an effect called desynchronization. This desynchronizing action of reticular stimulation is shown by the control trace in the middle of Fig. 6-6. The desynchronization continued long after the end of the stimulation. This desynchronization is similar to that produced by sensory stimulation or by a problem in Fig. 4-1. As was argued in Lecture 4, in the alpha rhythm the cortical neurones tend to beat together, probably by thalamo-cortical interaction (cf. Fig. 4-3). This rather fragile synchronization is disturbed when some other activating process is superimposed on the thalamo-cortical reverberation as in Figs. 4-1 and 6-6. In the upper trace of Fig. 6-6 olfactory stimulation had a very effective desynchronizing action. Recording from individual neurones shows that an increase in mean firing frequency also frequently accompanies the desynchronization.

In Fig. 6-6 both of the desynchronizing influences were almost abolished by a level of anaesthetic action that did not appreciably affect the alpha rhythm. As shown in the lowest trace of Fig. 6-6, doubling of the stimulus strength restored somewhat the desynchronizing influence of the reticular stimulation. The depressant action of the anaesthesia on the desynchronization suggests that there is

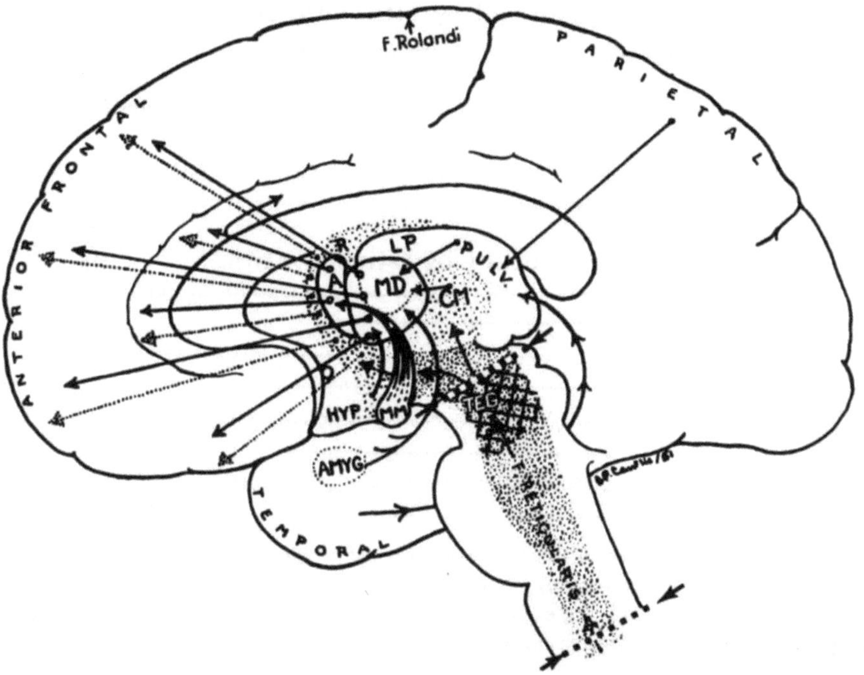

Fig. 6-5. Schematic representation of some of the principal afferent projections to frontal and anterior cingulate cortex (direct hypothalamic connections have been omitted). *A*, anterior nucleus of the thalamus; *MD*, nucleus medialis dorsalis; *CM*, centrum medianum; *MM*, mammillary body. The *stippled area* represents the ascending reticular system with unspecific projections indicated by the *dotted lines*. The hatched area labelled TEG is the midbrain reticular nucleus with ascending connections to the thalamic nuclei shown by *arrows*. Some connections of the amygdala (*AMYG*), Hypothalamus (*HYP*) and Mammillary body are shown (cf. Fig. 5-1). *TEG*, tegmentum; *LP*, lateral posterior thalamus; *PULV*, pulvinar. Dotted lines indicated by arrows show levels of sections. Penfield, W., Jasper, H.H.: Epilepsy and the functional anatomy of the human brain, pp. 896. Boston: Little, Brown & Co. 1954

a synaptic relay on the reticular pathway to the neocortex, and this indeed is the case. The synaptic relays are in the non-specific thalamic nuclei some of which are indicated in Fig. 6-5 (CM, LP, MD, A) and which in turn widely project to the neocortex, as is very inadequately indicated in Fig. 6-5. This wide thalamic projection is shown in Fig. 5-3, where the various thalamic nuclei are indicated by coded markings.

Inubushi et al. (1978a, 1978b) have recently made a very careful analytic study of the manner in which reticular stimulation evokes

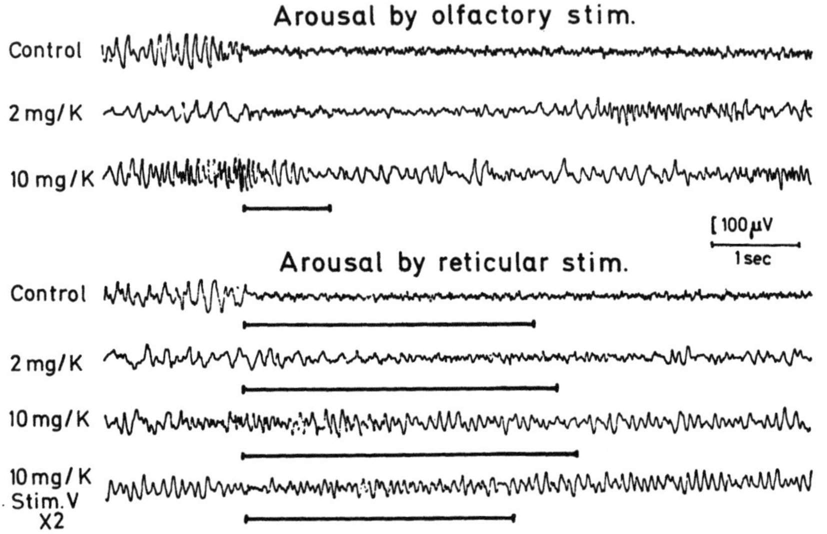

Fig. 6-6. Ink-written records of electrocorticogram of rabbit showing effect of pentobarbitone (Nembutal) (doses in mg/kg body wt.) upon EEG arousal induced by afferent stimulation (*above*) and direct excitation of brain stem reticular formation (*below*). Magoun, H.W.: The waking brain, pp. 135. Springfield, Ill.: Thomas 1958

arousal in the cortical neurones. By intracellular recording it is shown that there is an initial excitation of the neurones of lamina II (cf. Figs. 2-2 and 2-6) and later spread to the neurones of deeper layers by both excitatory and inhibitory synaptic action.

Normally there is always a background of activation of the cerebral cortex by the reticulo-thalamic pathway that maintains the cerebral cortex at a basal level of excitation. The stimulations in Fig. 6-6 were superimposed on that background. Abolition of the reticular influence by an appropriate lesion of the midbrain results in a great depression of the cerebral cortex. The EEG is reduced to that of a state of coma or stupor (traces 4 and 7 of Fig. 6-4) and the animal permanently remains in a state of deep sleep (Magoun, 1958). The site of the effective lesion is indicated by the cross-hatching in Fig. 6-5, on analogy with the lesions in a monkey brain (Magoun, 1958, Fig. 32). Arousal can still be evoked by stimulation of the activating pathway in the non-specific thalamus. The ascending sensory pathways shown diagrammatically in Figs. 5-4, 5-5 and 5-7 should still evoke cortical activity sufficient for arousal, but they would be severely damaged by the lesion shown in Fig. 6-5. It is to be noted that the paleospinothalamic pathway of Figs. 5-4 and 5-5 contributes to excitation of the reticular formation and so to the operation of the reticular activating system.

In investigating the effects of transverse sections of the neuraxis at different levels Bremer (1937, 1953, 1954) made the remarkable discovery that, when the section was made between the brain and the spinal cord (the lower transverse broken line of Fig. 6-5), the animal showed a normal EEG and a normal pattern of sleep and wakefulness. In contrast, when the section was made at the mesencephalic level (the upper transverse broken line of Fig. 6-5), there ensued a permanent condition resembling sleep. It is the same condition that has already been mentioned as occurring after destruction of the reticular activating system by the lesion indicated in Fig. 6-5. Thus Bremer's findings were related to the evidence for a reticular activating system by Moruzzi and Magoun in 1949. From all this evidence for a reticular activating system located in the mesencephalon there arose the concept that sleep was a passive process resulting from diminution of the wakefulness provided by the activating system. However, difficulties have arisen for this attractively simple explanation of sleep by deprivation, as will be discussed in the section on sleep.

6.3.3 Serotonergic Pathway

Figure 6-7 shows in the rat brain the nuclei of serotonergic nerve cells and the pathways that have been revealed by a fluorescence technique and also by an immunohistological method. This system is believed to be effective by serotonin as the synaptic transmitter. The raphe nuclei (principally B7 and B8 in Fig. 6-7) lie close to the midline, and, as shown, send axons rostrally along the medial forebrain bundle (cf. Fig. 5-1B) to be very widely distributed to the neocortex, as indicated in Fig. 6-7. As Phillis (1970) states, the accumulated evidence strongly suggests that serotonin (5-hydroxytryptamine) is a synaptic transmitter in the cerebral cortex. Iontophoretically applied serotonin depresses the excitability of cortical pyramidal cells, for example, in one experimental series, 89% of the 131

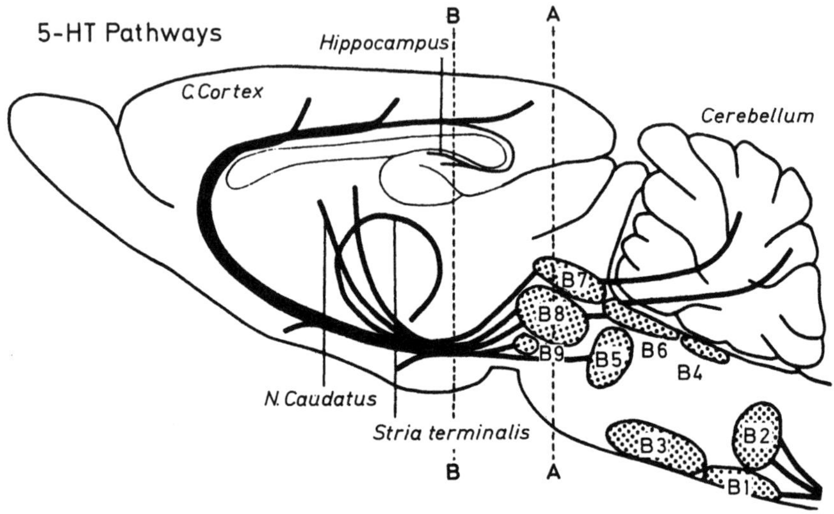

Fig. 6-7. Schematic diagram of the central serotonergic cell groups and projection in sagittal section from the rat brain: The nine major groups of serotonin cells in the rat are designated B_1 to B_9. BB and AA show levels of experimental sections McGeer, P.L., Eccles, J.C., McGeer, E.R.: Molecular neurobiology of the mammalian brain, pp. 644. New York: Plenum 1978

cortical neurones tested. It thus would seem that the serotonergic pathway is well fitted to be concerned in the active production of sleep, as will be described in the next section.

6.3.4 *Noradrenergic Pathway*

This pathway resembles in general form the serotonergic pathway, running parallel with it from its principal cell station, the locus coeruleus, which is approximately at the level of nucleus B6 in Fig. 6-7. There is considerable disagreement with respect to the action on cerebral neurones of the transmitter, noradrenaline (Phillis, 1970; Krnjević, 1974) that would be liberated from the terminals in the cerebral cortex. The arousal produced by intravenous injection may be secondary to excitatory action on the reticular activating system. Many cortical neurones are depressed by electrophoretically applied noradrenaline, but a delayed long lasting excitatory action has also been observed. Further reference to the action of the noradrenergic pathway will be made when dealing with the explanations of paradoxical sleep.

6.4 Sleep

6.4.1 Introduction

Sleep raises extraordinary problems for the brain–mind theorist no matter what philosophy is espoused. It also raises a host of neurobiological problems that are still without solution. It turns out that there are complex interacting systems in the brain stem, the thalamus, the basal ganglia and the limbic system which ultimately find expression on the neuronal machinery of the cerebral cortex. Furthermore sleep is not simply a passive process. There are also excitatory elements in the cortical responses, which overtly can give dream experiences. It is surprising that sleep cannot be regarded as a recuperative period for metabolic recovery of the exhausted waking brain. We have already seen that the oxygen consumption and circulation rate may be increased, and matching this is a level of neuronal discharge which may be above normal. A further difficulty is that most of the experiments have been on cats, which have a pattern of sleep behaviour very different from the human diurnal pattern.

6.4.2 Electrophysiology of Sleep

Figure 6-8 illustrates the stages of the EEG responses as sleep progressively deepens. In the upper trace there is the same excited condition of the cortex as in Figs. 4-1 and 4-2A, and the second trace is a good alpha wave response in the relaxed presleep stage. The third trace shows the slow wave stage of light sleep or quiet sleep with an irregular wave form at about 3 to 5/s. This is the kind of sleep through most of the night, but every 80 to 120 min there are periods of paradoxical or active sleep as indicated in the fourth trace of Fig. 6-8.

It is so named because the slow waves of quiet sleep are desynchronized to give a fine irregularity, much as in the excited state, on which are superimposed brief bursts of high frequency (12–15/s) called spindles, two of which are seen on the fourth trace of Fig. 6-8. At the same time there are horizontal rapid eye movements (REM), so it is often called REM sleep. There may be occasional body movements, but for most of the active or paradoxical sleep there is complete muscular relaxation, and there are dream episodes, as

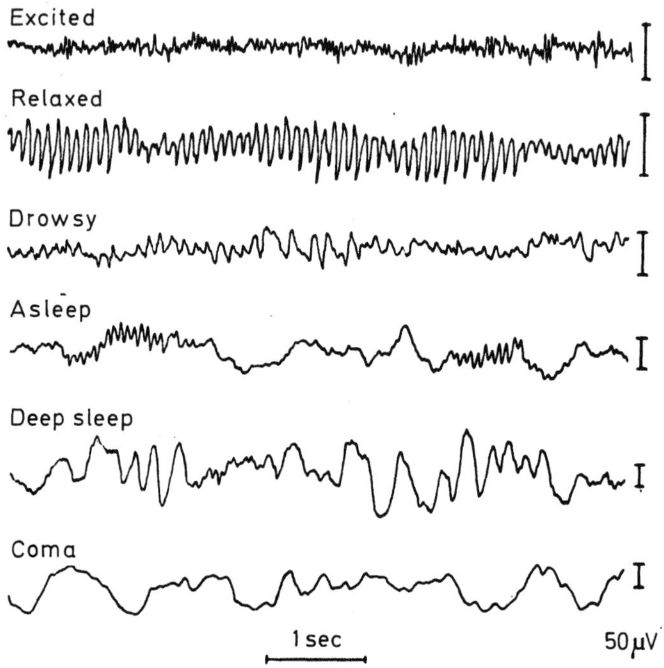

Fig. 6-8. Characteristic electroencephalograms during variations in states of consciousness and unconsciousness. Penfield, W., Jasper, H.H.: Epilepsy and the functional anatomy of the human brain, pp. 896. Boston: Little, Brown & Co. 1954

will be discussed later. The fifth trace of Fig. 6-8 is for extremely deep sleep with the slow delta waves. The three sleep states of the third, fourth and fifth traces in Fig. 6-8 represent progressively deeper sleep, as tested by the difficulty of arousal.

Studies of the neuronal activity during sleep have given insights into the generation of the EEG wave forms. For example, Evarts (1964, 1967) finds that the fairly regular rhythmic discharge of pyramidal tract neurones in the waking state (Fig. 6-9A) becomes broken into an irregular pattern with occasional bursts during slow wave or quiet sleep (Fig. 6-9B). With paradoxical or active sleep there are strong bursts of neuronal discharges standing out from virtual silences (Fig. 6-9C). Presumably in Fig. 6-9B there is some synchronization of the irregular burst responses in adjacent neurones to give the slow wave EEG of trace three in Fig. 6-8, while there is complete desynchronization of the responses of the individual neurones to

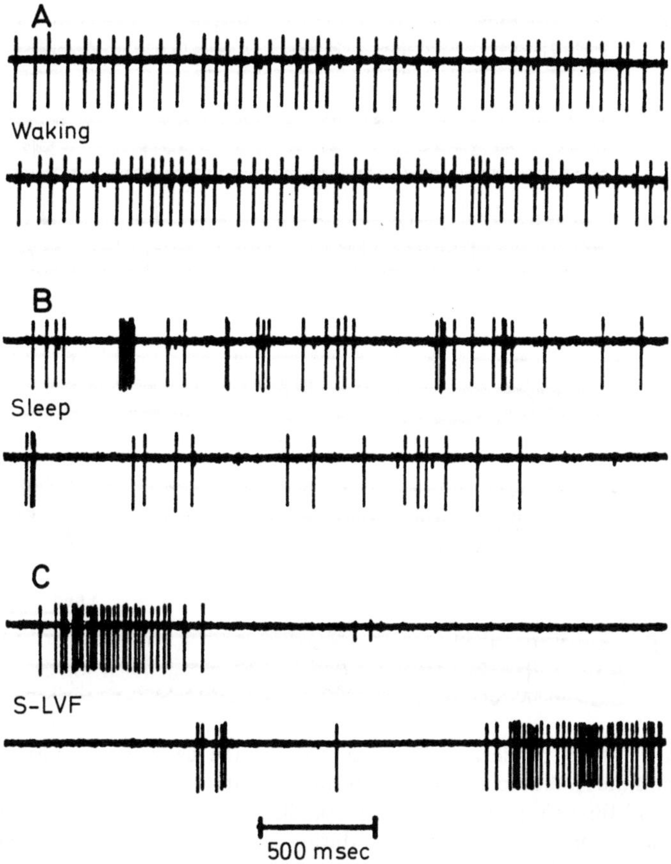

Fig. 6-9 A–C. Patterns of discharge of pyramidal tract neurones (*PTN*) during wakefulness and the two stages of sleep. Intact, unanaesthetized monkey. During wakefulness (two traces) the discharge is regular, without any tendency to clustered firing (**A**). During synchronized sleep (two traces) there are bursts interspersed with periods of relative inactivity (**B**). During sleep with low voltage fast activity (*S-LVF*) (two traces), burst duration increases, intervening periods of inactivity become longer, and discharge frequency rises (**C**). Evarts, E.V.: Temporal patterns of discharge of pyramidal tract neurons during sleep and waking in the monkey. J. Neurophysiol. *27*, 152–171 (1964)

give the EEG of the fourth trace of Fig. 6-8 between the spindles, which signal some degree of synchronization.

Synchronization can be tested by simultaneous recording from two or more neurones (Creutzfeldt and Jung, 1961). It is to be noted that in Fig. 6-9 there are two traces from *the same* neurone

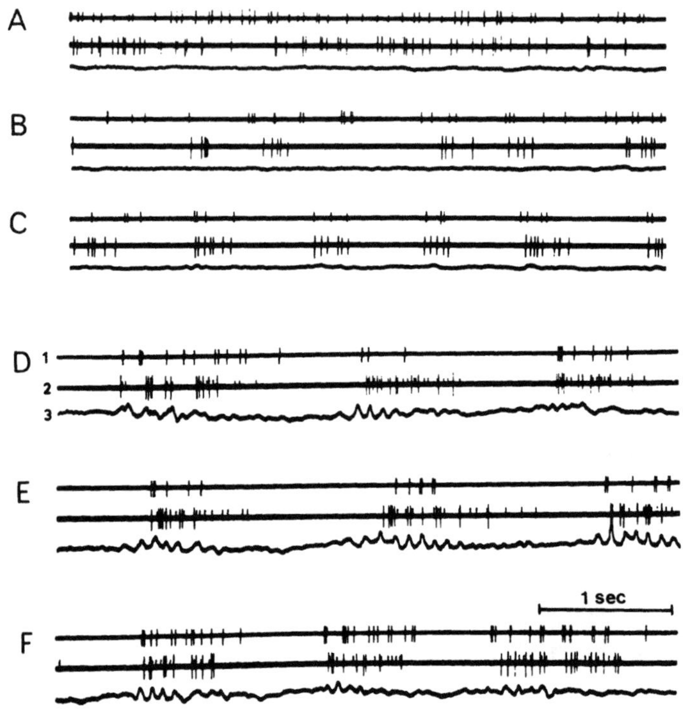

Fig. 6-10 A–F. Onset of sleep in two neurones of the cat's motor cortex. After a brief aroused state **A**, the records **B** and **C** were taken at 2-s intervals during progressive drowsiness and final sleep showing 'sleep spindles' of 12–13 per s in the EEG. Both neurones change their discharge pattern during sleep to a periodic grouped activity with longer pauses. They slow their average frequencies from around 12 per s in arousal to 3–9 per s in falling asleep. **D**, **E** and **F** Periodic grouped discharges of three neurones and EEG of the cat's motor cortex during spindle activity in sleep. Neurones 1 and 2 are picked up by two microelectrodes 1.5 mm apart. The third neurone appears with lower amplitude spikes in the background of channel 2. The neuronal group discharges mostly occur simultaneously with the 10 per s spindles of the EEG (trace 3) although no strict time relation with individual waves can be found. Creutzfeldt, O., Jung, R.: Neuronal discharge in the cat's motor cortex during sleep and arousal. In: The nature of sleep. Wolstenholme, G.E.W., O'Connor, M. (eds.), pp. 129–170. London: Churchill 1961

in each series. The two traces in Fig. 6-10A show the desynchronized responses of two neurones in the waking state as in Fig. 6-9A. At the onset of sleep (Fig. 6-10B) there are occasional burst discharges and some synchronization corresponding to the responses of Fig. 6-9B. With deeper sleep in Fig. 6-10C there are synchronous bursts of the two neurones. Each of these bursts gives a sleep spindle of up to a second duration (cf. trace 4 of Fig. 6-8). In Fig. 6-10D, E and F, there are recordings from three neurones, there being a small and a large spike series on the middle trace. The lowest trace is the EEG showing clear spindle bursts that are approximately synchronous with the synchronized spike bursts. Each wave of the spindle is generated by the summing of many neuronal discharges in approximate synchrony.

During quiet sleep the total number of neuronal discharges tends to be diminished, but with paradoxical sleep it is as high as in a normal resting state (Evarts, 1964). In the monkey the large motor pyramidal cells fire slowly in the waking, resting state and tend to fire faster during light sleep, while those firing rapidly (the small cells) tend to slow in frequency during sleep, as is indicated in Fig. 6-11A (Evarts, 1967). There is thus much more uniformity in the mean neuronal frequencies during sleep.

These same frequency changes with sleep also occur with non-pyramidal tract cells of the motor cortex, and with other cortical neurones. For example in Fig. 11B with visual cortical cells the mean rate of firing was about 8/s during waking (W) and during quiet sleep (S), but during paradoxical sleep (S-LVF) the mean frequency, 15/s, was almost as high as during visual stimulation in the awake cat (W-VIS, 16/s) (Evarts, 1967). The values are for overall means and individual neurones may change considerably in their frequency at different levels of sleep. The burst discharges in paradoxical sleep and to a lesser extent in quiet sleep (Fig. 6-9) result in a great increase in the short interspike intervals, those less than 10 ms; and Evarts (1967) suggests that this disturbance of temporal pattern may be related to the unconsciousness of sleep.

Noda and Adey (1970, 1973) investigated the firing patterns of single neurones in the association cortex of the cat. As with the motor and visual areas the firing rates decreased as a rule in sleep, but increased in paradoxical sleep to high rates. With deep sleep and spindle sleep there was a high correlation between the temporal patterns of two units recorded simultaneously (cf. Fig. 6-10C, E, F). However in arousal and REM sleep there was no correlation.

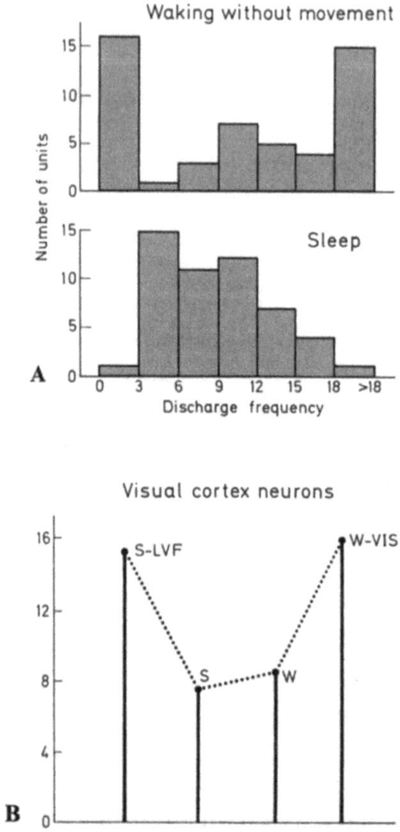

Fig. 6-11. A Distribution of discharge frequencies in the sample of 51 PTNs during waking in the absence of movement (W) and sleep with EEG slow waves (S). During W, 16 of 51 units had discharge frequencies less than 3/s, whereas only one of the 51 units had such a low discharge frequency in S. During W there were also many very active units, 15 of 51 units having discharge frequencies over 18/s. With S, only one of the 51 units had a discharge frequency above 18/s. **B** Discharge frequencies of visual cortex neurones. During quiet sleep with slow waves (S) the mean frequency is a little slower than during waking (W), but it is much higher during paradoxical sleep (S-LVF), being almost as high as during visual stimulation (W-VIS). Evarts, E.V.: Unit activity in sleep and wakefulness. In: The neurosciences. Quarton, G.C., Melnechuk, T., Schmitt, F.O. (eds.), pp. 545–556. New York: Rockefeller University Press 1967

6.4.3 Explanations of the Phenomena of Sleep

We have already discussed the role of the reticular activating system in arousal and the deep depression that ensued when this activating influence was removed. There has been brief reference to the simple explanation of sleep as a passive process that was derived therefrom (Bremer, 1954). It is remarkable that after section of the long ascending and descending tracts in the mesencephalon, but with sparing of the reticular system, the animals can be aroused and show no abnormalities of the sleep–waking cycle. However, Bremer did not propose a simple one-way action from reticular system to cortex. There is much evidence for a cortico-reticular pathway, so there is opportunity for cortico-reticular-cortico etc. reverberatory circuits that can maintain arousal, as suggested by Moruzzi and associates for the reticular formation at lower levels of the brain stem.

Much of the research has been carried out on the cat which is much more sleep prone (about 75%) than humans. An important discovery was made by Jouvet (1972, 1974) that, besides the passive process of suppression of the reticular activating system, sleep is also an active inhibitory process, being brought about by the raphe nuclei and the serotonergic pathway (Fig. 6-7).

Jouvet has shown that removal of 80% to 90% of the raphe nucleus converts cats into insomniacs with reduction of sleep to 10%–15%. This alert state of the cat goes on permanently. Injection of the precursor of serotonin does not diminish the arousal, presumably because the serotonin synthesizing mechanism is destroyed. It should be noted that serotonin injected into the blood stream is ineffective because access to the brain is blocked by the blood–brain barrier.

Injection into normal cats of a drug that prevents the production of serotonin in the raphe nucleus results after some 20 h in an abrupt decrease in both the slow wave and the paradoxical sleep, so that after 30 to 40 h the cat has an insomnia even more severe than after excision of the raphe nuclei. However this can be temporarily reversed (for 6–8 h) to a normal sleep pattern by injection of a drug that is a precursor of serotonin by a process that is not blocked by the background injection (Jouvet, 1974). So Jouvet (1974) concludes: 'These investigations show that sleep mechanisms can be manipulated by interfering only with the synthesis of 5 HT (Serotonin)'.

A much more severe problem is raised by the question: What triggers the serotonin neurones? The waking mechanism is almost indefatigable because a subtotal insomnia may be obtained after the raphe lesion. Thus the serotonin mechanism has to be in action during the whole sleep cycle, and in fact, there is a high serotonin level in the brain throughout the whole sleep cycle. Jouvet (1974) suspects that the answer may be obtained from studies on the nuclei at lower brain stem levels that would activate the raphe nuclei.

Linked with the inhibitory action of serotonin on cortical neurones is the effect of a related system, the noradrenergic as referred to above. Jouvet (1974) suggests that this system is responsible for the arousal effects of paradoxical sleep, which can be seen in trace four of Fig. 6-8 and also for arousal at the end of a sleep period. Furthermore, Jouvet postulates that the serotonin system may be responsible for activating the noradrenaline secretion from the locus coeruleus.

Still further complications relate to the postulated controlling influence on the sleep cycle by a bulbar system, possibly the nucleus of the solitary tract (Bremer, 1977). Ascending fibres inhibit the reticular activating system. This bulbar system is superimposed on the serotonin system (cf. Fig. 6-7). But many more interacting systems are proposed. A useful summary has been given by Fuxe (1972) for the serotonin and noradrenaline systems. Furthermore, Bremer (1977) gives evidence for a sleep initiating centre in the preoptic area in association with the hypothalamus, there being reciprocal tonic inhibitory action with the reticular activating system.

The normal 24-h cycle of sleep and waking is built up by many factors – day and night, patterns of meals and of work. But there are also important intrinsic controls. This can be revealed by experiments 'without time' or 'free running' in caves or in bunkers (cf. Bert et al., 1978). Under these conditions it was surprising that most subjects held close to a 24-h cycle that ran synchronously with body temperature changes. But some moved even to a 48-h cycle with 34 h awake and 14 asleep, still thinking that they were on a normal 24 h cycle! The internal control of sleep–waking cycles is very complex involving hypothalamic control of pituitary secretions. Evidently it is a disturbance of these controls that results in the unpleasant experiences of 'jet-lag'.

Narcolepsy and hypersomnia present remarkable examples of disorders in human sleep. There are excellent reviews by Roth (1976,

1978), to which reference should be made for a full account of the clinical details. In idiopathic narcolepsy there are states of irresistible 'sleep' of short duration, exceptionally up to 10 to 15 min, in which the patient loses muscle control and may fall to the ground. There is inability to move and to speak. Consciousness is usually preserved, but it may be clouded. In hypersomnia there are during the day states of excessive sleepiness that result in sleep of one to several hours duration in which consciousness is lost as in normal sleep, and there even may be associated dreams. It would seem probable that disorders of the serotonin mechanism are responsible for these phenomena.

6.4.4 Dreams

A remarkable advance was made when it was discovered that dream states were usually signalled by rapid eye movements (REM) and by the emerging out of a spindle stage of the EEG, but to a stage of deeper sleep as shown by the difficulty of awakening. When aroused during or shortly after the termination of REMs, there was almost always the report of having dreamed (Kleitman, 1960, 1961, 1963). The recall of the dream was best when the subjects were awakened during the REM. Detailed recall was given by 46 out of 54 awakenings during REM, but only fragmentary recall in 9 out of 11 awakenings within 5 min of the cessation of the REMs, and complete forgetfulness in 25 out of 26 awakenings at 10 or more min after the REM. However, some subjects do not have a dream association with REM sleep. Sometimes, when wakened up in the quiet sleep phase, they report being wakened from a dream.

It is suggested that the REMs are due to the eye scanning the dream images which mostly move horizontally. As tested by awakening, it is found that everybody dreams several times every night, but of course, most dream experiences are lost unless recalled by an awakening during or soon after the dream. Dream episodes last for 10 to 35 min followed by an interval of about 90 min so that there are about four episodes a night with a total dream time of 1 to 2 h. If dreaming is curtailed by 75% to 80% by awakening subjects as soon as the REMs appear and this is continued for several nights, the subjects become anxious and irritable with an increased appetite. At the end of this regime there are several nights of increased dreaming. This does not occur when the awakenings did not interrupt dreams. It is therefore considered that dreaming has some salutary influence, being a kind of self-applied psycho-analysis.

6.4.5 Hallucinations

We can validate our normal perceptual experiences by intersubjective communication for the sights and sounds that we are immersed in. With hallucinations this validation breaks down. A person can hear voices or see some 'object' as a UFO or a ghost and these experiences are not shared by companions. Hallucinations are a familiar dramatic device. In some mental disorders, such as schizophrenia, hallucinations are very common. Julian Jaynes (1976) proposes that until the forging of critical thinking in Greek philosophy mankind lived in a world dominated by hallucinations, particularly of heard voices of the oracles or with the priests in the ziggurats. It would be an extension of dream-like states into the waking existence. Of course we can all experience such states in the transition from sleep to waking, where we are in part awake and in part still in our dream.

Two days ago the BBC (May 24) showed on TV an extraordinary case of hallucination. A lady described her experiences with reincarnated composers who had been dictating their new compositions to her for many years. Liszt came to her first in childhood, then again many years later when she recognized him. Later Beethoven appeared and in his reincarnation was young and not deaf. She has now written down a large number of compositions from dictation by Lizst, Beethoven, Chopin and other great composers. She actually continued with her most recent composition during the interview, being quite unperturbed by the occasion. A pianist played quite interesting music alleged to be delivered by Chopin! There is to be a full-scale recital of her deliverances in London next week-end. You will agree that there are many mysteries in the mind and in the brain!

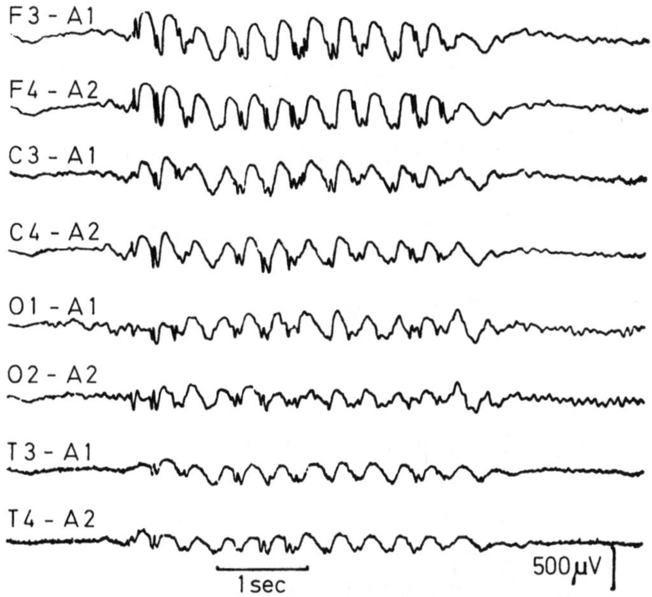

Fig. 6-12. A small petit mal seizure with electroencephalogram showing the involvement of all head regions, even during a minor attack without loss of consciousness. Penfield, W., Jasper, H.H.: Epilepsy and the functional anatomy of the human brain, pp. 896. Boston: Little, Brown & Co. 1954

6.5 Convulsions

The electroencephalogram reveals that during epileptic seizures there are generated strong rhythmic waves at about 3/s. In the mild and brief seizures called petit mal only a limited area of the cerebral cortex is involved for a brief time. In Fig. 6-12 the EEG shows the considerable extent of a typical wave form lasting for a few seconds. The subject did not lose consciousness, though some mental confusion is usual.

A more severe petit mal seizure is illustrated in Fig. 6-13. The seizure lasted for more than 10 s and high voltage waves at about 3/s were recorded from both frontal and occipital areas on both sides. As would be expected from this widespread severe involvement, consciousness was lost. It is to be noted that from the frontal origin it takes several seconds to involve the occipital lobe. In this patient the hippocampus also participated in the seizure. Figure 6-4 trace 3 also illustrates a recording of a petit mal attack when consciousness

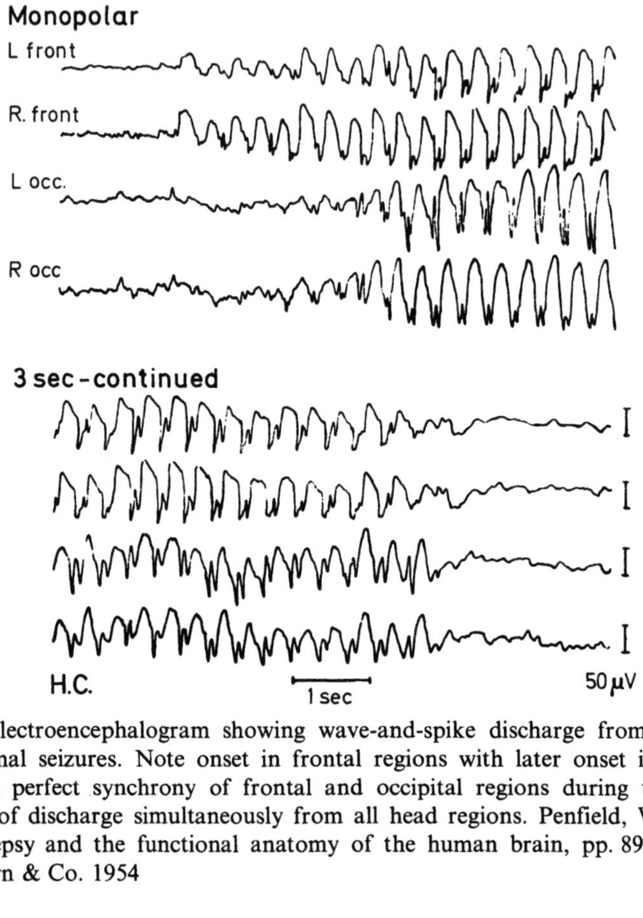

Fig. 6-13. Electroencephalogram showing wave-and-spike discharge from a patient with petit mal seizures. Note onset in frontal regions with later onset in occipital regions and perfect synchrony of frontal and occipital regions during the attack, with arrest of discharge simultaneously from all head regions. Penfield, W., Jasper, H.H.: Epilepsy and the functional anatomy of the human brain, pp. 896, Boston: Little, Brown & Co. 1954

was lost. The onset of a major seizure with a faster rhythmic beat is shown in trace 5 of Fig. 6-4.

Jasper (1969) states that the neuronal basis of the epileptic wave must be a large aggregate of neurones firing at the high frequency of 50 to 500/s with synchronization of adjacent neurones. All finely organized patterns of activity are abolished by the strongly driven discharge. In experimental epilepsy produced by cortical injury, subcortical structures such as the thalamus also participate in the epileptic discharge. However the spread of discharge in the cortex is not dependent on these subcortical structures. It occurs by the association and commissural pathways between the modules (cf. Figs. 2-5 and

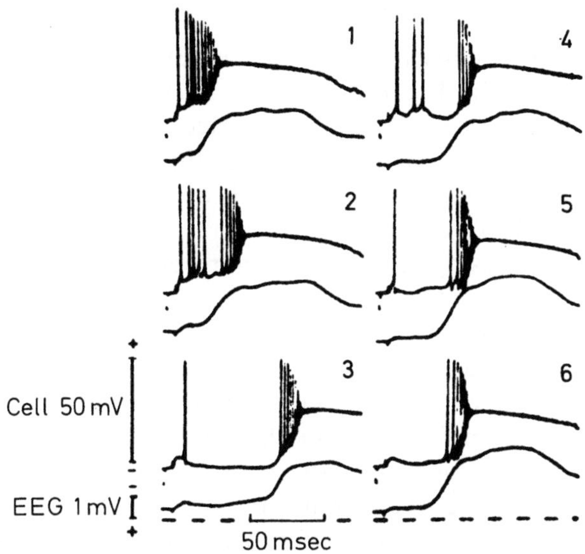

Fig. 6-14. Evoked seizures in a penicillin focus in the anterior sigmoid gyrus of the cat. Stimulation of the VL thalamic nucleus every 2 s. Different stimulus intensities (arbitrary units): 20 in 1, 15 in 2, 7 in 3, 10 in 4–6. Upper traces are intracellular recordings, lower the EEG from the surface of the cerebrum nearby. Matsumoto, H., Ayala, G.F., Gumnit, R.J.: Neuronal behaviour and triggering mechanism in cortical epileptic focus. J. Neurophysiol. *32*, 688–703 (1969)

2-7). Intracellular recording from pyramidal cells (Fig. 6-14) shows that each surface potential wave of a seizure is generated by a strong synaptic depolarization of the pyramidal cell that generates a high frequency discharge until this is turned off by the extreme depolarization that then slowly declines. It can be assumed that there is a tremendous build up of excitatory power by neuronal interaction within the module (cf. Fig. 2-6) and also by inputs from other modules. It appears that the normal inhibitory control is in abeyance, and this would be at least a contributory factor to the propensity for epileptic seizures (cf. Ajmone-Marsan, 1969; Creutzfeldt, 1969).

Matching the intense neuronal activity of seizures it has been reported above that the metabolism and blood flow increase up to several times in the area of the seizure.

6.6 Coma and Vigil Coma

Coma is a deeply unconscious state. The EEG may show large slow irregular waves as in Fig. 6-4, trace 4, or in Fig. 6-8, trace 6, or it may be almost flat as in Fig. 6-4, trace 6. Correspondingly the neuronal discharges are greatly depressed, there being usually neuronal silence, and the metabolism is much depressed – down to 58% to 49% of normal in Table 6-1.

The most distressing kind of coma results from brain injuries sustained in accidents. The cerebral cortex may be virtually undamaged, the lesion being in the brain stem and causing a destruction of the reticular activating system, as is indicated in Fig. 6-5. This 'vigil coma', as it is called, may continue for months or even years with the patient immobile and unable even to move the eyes or make any sign of recognition. In some fortunate cases the coma may pass off after some weeks, but almost never after some months. Hassler (1978) has reported considerable success in these prolonged cases by attempting to substitute for the reticular activating system. After a course of prolonged stimulation – 1 h per day for 20 days – of the unspecific thalamic nuclei, CM, MD and A of Fig 6-5, the patients displayed some recovery from a coma of several months with complete immobility.

6.7 Anaesthesia

The unconsciousness of anaesthesia is easily explained. There is a great reduction of the neuronal activity of the cerebral cortex. The brain metabolism was reduced to 64% in Table 6-1. The EEG was greatly depressed in Fig. 6-4, trace 2. In Fig. 6-15 (Creutzfeldt, 1975) the spontaneous discharge rate of cortical neurones was progressively depressed by a continuous injection of a barbiturate, though with a low dose there may be a transient increase in frequency. Thereafter all neurones were depressed and most were silenced. Correspondingly the EEG is shown in the vertical traces above to be depressed eventually to a flat record.

Noda and Adey (1973) observed that in anaesthesia the firing rate of neurones of the association cortex was reduced to 30% to 50% of that in slow wave sleep, being far below that of normal wakefulness and paradoxical sleep. With slow wave sleep (deep) and anaesthesia there were much the same firing patterns of the neurones with high correlation in the firing of adjacent neurones.

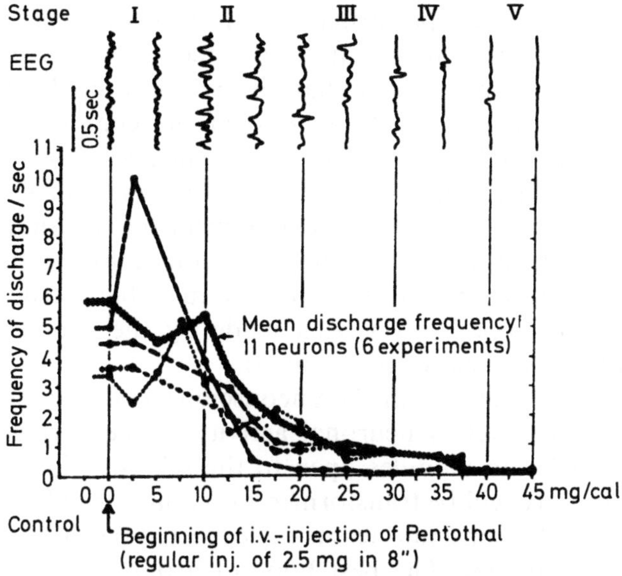

Fig. 6-15. The effect of a barbiturate on the spontaneous discharge rate of cortical neurones. 2.5 mg thiopentone (Pentothal) were injected every 8 s (abscissa). The mean discharge frequency of a sample of 11 cortical neurones from different experiments is shown as a *heavy line* and the changes of four typical single neurones by the thinner, *interrupted lines.* The EEG recorded at the different stages is shown on top of the graph with the correspondent numbers of anaesthetic states. Note, that after 10 mg pentobarbitone (Nembutal) i.v. per animal some neurones may be activated, others depressed, but that beyond this dose all units show decreased discharge rate. Weight of animals 2.0–2.2 kg. Creutzfeldt, O.D.: Neurophysiological correlates of different functional states of the brain. In: Brain work: The coupling of function, metabolism and blood flow in the brain. Ingvar, D.H., Lassen, N.A. (eds.), pp. 21–46. Copenhagen: Munksgaard 1975

6.8 Discussion on Brain Activity and Levels of Consciousness

In this lecture we have surveyed the levels of cerebral activity that characterize the extremes from coma to convulsions with in between, the physiological levels of normal waking and sleeping. It is a challenge to brain–mind theories to relate the observed cerebral events to the levels of consciousness. But meanwhile, it is important to interpret the various brain activities in terms of the modular operation of the cerebral cortex (cf. Fig. 2-6) as described in Lecture 2. There we considered the evidence that the modules act as operative units and that all brain performance is in terms of spatio-temporal patterns of modular activities as illustrated in Figs. 2-5, 2-7, 2-8, 2-9 and 2-10. It is postulated that the relationship between mind and brain is a reciprocal action between the conscious mind and the open modules (cf. Figs. 1-7 and 2-10) in accord with attention and intention as described in Lecture 4. My present task is to show how the hypothesis of dualist-interactionism (cf. Lectures 1 and 2) can provide explanations for the levels of consciousness that characterize the observed cerebral activities. It may be noted in passing that no such attempt has yet been made by the exponents of the various materialist theories of the mind.

There is no problem with the unconsciousness of coma and deep anaesthesia. The level of neuronal activity in the cortex is so low that there will be no modules open to the self-conscious mind. For example, Fig. 2-10 will be transformed into a uniformly black ensemble of modules. Unconsciousness is the expected result of this closure of almost all modules. The best hope of arousal from coma is to stimulate the reticular activating system and its thalamo-cortical pathways. The ordinary sensory pathways seem to be much less effective, perhaps because they primarily activate the specific thalamic nuclei that do not have a widespread dispersal to the cerebral cortex (cf. Fig. 5-3).

In seizures we have to recognize that whole modules are powerfully activated with neurones firing intensively to exhaustion, as is illustrated for the intracellular recording in Fig. 6-14. After recovery in a few hundred milliseconds the cycle is repeated and so on for the whole duration of the seizure (cf. Figs. 6-12 and 6-13). It has been suggested that the cycle of about 3/s may be in part determined by reverberatory loops either to the thalamic nuclei or entirely within the cortex. It can be assumed that modules in these alternating states of intense firing and exhaustion cannot participate in the gentle two-way influences (cf. Fig. 1-7) that characterizes the brain–mind interaction, as described in Lectures 2, 3 and 4 and Eccles, 1979a, Lecture 10. So again in the diagram of Fig. 2-10 those brain regions involved in the seizures will be assemblages of closed modules labelled in black. However, in minor petit mal seizures (cf. Fig. 6-12) even the areas of the seizure were only partly involved, as indicated by the low voltages of the waves. Also much of the cortex is not invaded by these seizures. As a consequence there would be considerable areas of the cortex with the normal pattern of open and closed modules as in Fig. 2-10, brain–mind liaison being thus preserved. However, the conscious patient may experience mental confusion as a consequence of the disordered activity in large regions of the brain.

With major seizures (grand mal) consciousness is not lost until there has been invasion of a considerable area of the cortex perhaps up to one-half. In this connection it is of interest that patients do not lose consciousness when a whole right hemisphere is removed under local anaesthesia (Austin et al., 1972). In fact they can continue in conversation through the whole operation!

Often disordered sensory experiences (auras) precede the loss of consciousness in an epileptic seizure. This would be expected to occur when there is partial activation of perceptual areas, just as occurs with electrical stimulation.

The unconsciousness of sleep presents a much more perplexing problem. Neuronal discharges are somewhat reduced in slow wave sleep (Evarts, 1964, 1967) and there is a tendency for occasional bursts (Fig. 6-9 B). The small cortical cells that normally are very active tend to slow considerably during slow wave sleep (Fig. 6-11 A), while larger neurones increase their firing rate. It is significant that the alpha rhythm is lost (cf. Fig. 6-8) with the onset of slow wave or quiet sleep. It would appear that the reverberating thalamo-cortical circuits have been decreased in their effectiveness. This can be attributed to the strong inhibitory influence of the serotonergic pathway from the raphe nuclei to the thalamus, which in addition is depressed as a consequence of the depressed reticular activating system. There does not seem to be a sufficient change in patterns of discharge of cortical neurones to account for the unconsciousness of quiet sleep. The only suggestions that have been made attribute the effect to the alteration of the temporal pattern of neuronal discharges, particularly the tendency for burst discharges (cf. Fig. 6-9 B) (Evarts, 1967; Creutzfeldt, 1975; Schlag, 1974).

There is a necessity for exploration of the neuronal patterns of firing within the module in order to discover the reason for failure of the two-way interaction of Fig. 1-7. Already Inubushi et al., (1978 a, 1978 b) have reported a detailed study of the patterns of involvement of neurones in the different laminae of the cerebral cortex that occurs during arousal.

With paradoxical sleep the deeper unconsciousness is probably attributable to the bursting activity of neurones that often are synchronized (Figs. 6-9C and 6-10D, E, F) to give the sleep spindles. Despite the higher average frequency of neuronal discharge this disordered pattern may very effectively block brain–mind interaction. A further problem arises in attempting to explain the abnormal conscious experiences of dreams. Here there is a disordered consciousness divorced from the normal controls and rationalizations of the waking state that would be exercized across the brain–mind frontier of Fig. 1-7.

We have now to confront an intractable problem as we consider why it is necessary to sleep. It is certainly not a time for metabolic recuperation of exhausted neurones. There is during sleep no neuronal rest, and even an increase (cf. Figs. 6-9, 6-10, 6-11). Recuperation is not needed for the cerebral activities concerned with respiration or cardiovascular control. Moruzzi (1966a, 1966b) suggested that the slow plastic processes concerned in consolidation of memory (that will be described in the next lecture) may require long recuperative periods and that sleep is a device for having these more subtle regions of the cortex recuperate together, rather than having us half asleep and half awake all the time!

In line with this suggestion I would propose that during sleep there is a reconstitution of neuronal connectivities so that some become consolidated to give long-term memory. But this consolidation would only happen for a minute fraction of the neuronal events of the waking day. The immense barrage of sensory input and the immense internally generated neuronal activities have to be dissolved so that after the sleep the brain is reconstituted in readiness for the next day of intensive handling of data and selection of what is to be consolidated in long-term memory. So, as proposed by Moruzzi, sleep would be a time for the renewal of plastic processes.

Lecture 7

World 3, Remembrance and Creativity

Résumé

The *World 3 of objective knowledge* is recognized according to Popper as the world of culture and civilization. There are two components of World 3: that coded on artefacts such as books, pictures, tapes etc. and that coded in the memory stores of the brain. World 3 thus comprises the collective stores of all human creativity.

The relationship of *World 3* to the brain is accomplished by the lines of communication to and from the highest level of cerebral activities. In that way the conscious self can receive from World 3 by means of the liaison brain and can participate in creativity with respect to World 3.

In answering the criticisms of philosophers it is emphasized that World 3 has an objective existence. This is discussed in respect both to a scientific discipline such as chemistry and to creations in the plastic arts and in music. It is important to recognize that knowledge of World 3 is not inherited genetically even to the slightest extent. However, the brain is built by genetic instructions with the propensity to participate in the World 3 of culture.

There is an account of further developments of the theory of *cognitive learning* described in Lecture 9 of the preceding Gifford series. Special reference is made to the modular structure of the cerebral cortex. Diagrams are developed that show how the postulated synaptic hypertrophy could lead to changes in the spatio-temporal patterns of modular activity. The aim of this enterprise is to develop models that could account for the laying down of specific memory traces and that participate in the recovery of memories when spatio-temporal patterns of modules are replayed much as in the initial experiences to be remembered.

Finally there is a speculative account of the way in which creative insights can be attained and of the possible brain–mind interactions that are related thereto.

7.1 Introduction

The theme of this lecture is twined around three closely related topics, so closely related that I had difficulty in deciding the order of their presentation. World 3 can be understood as the cultural heritage of mankind. It is the product of cultural evolution (as described in Lecture 6 of Eccles, 1979a) and each one of us has become a civilized human being by immersion in the World 3 of our culture, as described in Lecture 7 of Eccles (1979a). This has been accomplished by learning and remembrance. Finally World 3 is constituted from the creativity of countless individuals from prehistoric times up to the present, and each of us has the challenge to add to World 3 by our own creative efforts. The close interrelationship of these three components of my lecture is apparent.

I have found Popper's account of World 3 of transcendent importance. It was originally expressed in two lectures (Popper, 1968a, 1968b) that were assembled in his book *Objective Knowledge* (Popper, 1972). For me it had solved problems relating to the nature and status of science that had long been of concern. But most philosophers, as for example at the World Congress of Philosophy at Düsseldorf in 1978, have been destructively critical, regarding it as a piece of gratuitous metaphysics. One suspects that they have a 'hang up' over Plato's third world philosophy. They do not of course accept it, but are suspicious of any philosophy that adopts this term, even though in a completely different connotation. Because of this widespread criticism I am here presenting a further account of Popper's World 3 philosophy together with a critical evaluation of the objections raised by philosophers, both at Düsseldorf and in the reviews of our book (Popper and Eccles, 1977) in which World 3 was treated at length in Chaps. P2 and P3 and in Part III, where there was extensive discussion in Dialogues III and X. Recently Dawkins (1976) has proposed a theory of 'memes' which corresponds to some extent to World 3. Apparently he did not know of Popper's publications up to 8 years earlier.

7.2 World 3

7.2.1 Initial Statement

Plato transcended the duality of body and mind by proposing a third world of forms or ideas which were true and eternal and which provided ultimate explanations, for example of beauty as having some share in the idea of absolute beauty. It was a theory of essences that fed-back on us, but which were divine and beyond human control. It was an attractive idealism that has exercised a great influence through the subsequent history of philosophy. There have been many variants on Plato's third world theme by philosophers from the Stoics to Plotinus to Hegel and in more recent times to Bolzano and Frege, as has been described by Popper (1968a, 1968b). Then, recently, Popper has developed a three-world philosophy, building on the two worlds of matter and mind a third world of objective ideas, with components partly enumerated in the third column of Fig. 7-1. This World 3 of Popper is fundamentally different from the third world of Plato in that it is a human and not a divine creation.

> It is the world of intelligibles, or of *ideas in the objective sense*, it is the world of possible objects of thought: the world of theories in themselves, and their

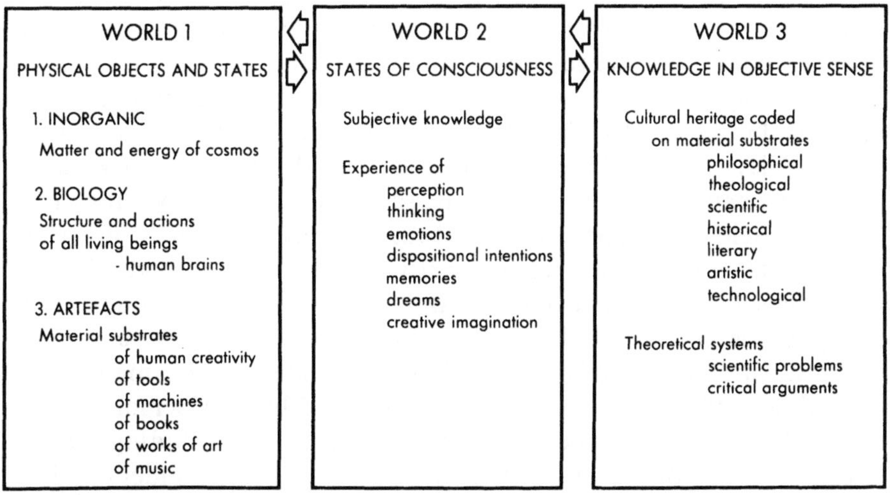

Fig. 7-1. Tabular representation of the three worlds that comprise all existents and all experiences, as defined by Popper. Full explanation in text.

logical relations; of arguments in themselves; and of the problem situations in themselves (Popper, 1968b).

Figure 7-1 represents an attempt to give a diagrammatic expression of the postulated contents of the three worlds, as proposed by Popper, and that were briefly referred to in Lecture 1. World 1 is the world of physical objects and states including all of biology, even human brains. It is important that it also comprises the material substratum of all human creativity, that is of all World 3 objects. World 2 is the world of states of consciousness or mental states. For present purposes we need not raise the issue of animal consciousness. As has already been discussed in Lectures 1 and 2, World 2 is our private subjective experience, that is known in others by inference from symbolic communications. Its content is shown in the second column of Fig. 7-1, and more adequately in Fig. 1-7.

By contrast, World 3 is the world of knowledge in the objective sense, and as such has an extremely wide range of contents. In Fig. 7-1 there is an abbreviated list. For example it comprises the expressions of scientific, literary and artistic ideas that have been preserved in codified form in libraries, in museums and in all records of human culture. In their material composition of paper and ink, books are in World 1, but the knowledge encoded in the print is in World 3, and similarly for pictures and all other artefacts. Most important components of World 3 are the theoretical systems comprising scientific problems and the critical arguments generated by discussion of these problems. In summary it can be stated that World 3 comprises the records of the intellectual efforts of all mankind through all ages up to the present – what we may call the cultural heritage. As Popper states (Popper and Eccles, 1977), World 3 is: 'the world of the products of the human mind, such as stories, explanatory myths, tools, scientific theories (whether true or false), scientific problems, social institutions and works of art'.

Thanks to the sum total of creative efforts of mankind World 3 is immensely rich and extensive and during a whole lifetime it is not possible for one individual to do more than know a minute fraction, and only rarely can a significant addition be made. The central task of the humanities is to understand human persons, while that of the natural sciences is to understand nature including human beings as part of nature. Since in both cases this understanding achieves expression in language, both can be regarded as branches of literature, and both have an honoured place in World 3.

7.2.2 *Objective Existence of World 3*

Popper has been at great pains to establish the objective existence of World 3. It is in this respect that his critics have failed to appreciate the convincing character of his demonstration of objectivity. Thus he writes (Popper, 1968a):

> What I regard as the most important point is not the sheer autonomy and anonymity of World 3, or the admittedly very important point that we always owe almost everything to our predecessors and to the tradition which they created; that we thus owe to World 3 especially our rationality – that is, our subjective mind, the practice of critical and self-critical ways of thinking, and the corresponding dispositions. More important than all this, I suggest, is the relation between ourselves and our work, and what can be gained for us from this relation.
> One of the main functions of World 2 is to grasp the objects of World 3. This is something that we all do; it is the essential part of being human to learn a language and this means, essentially, to learn to grasp *objective thought contents* ... in so far as language contains information, in so far as it says or describes anything or conveys any meaning or any significant message which may entail another, or agree or clash with another, it belongs to World 3. *Theories, or propositions, or statements are the most important World 3 linguistic entities.* (Popper, 1968b)

Popper emphasizes the importance of World 3 in our attempt to understand the scientific enterprise. This he illustrates by an example (Popper and Eccles, 1977, Section 11)

> The production of a scientific theory; its critical discussion; its tentative acceptance, and its application which may change the face of the earth, and thus of World 1. The productive scientist as a rule starts with a *problem*. He will try to understand the problem. This is usually a lengthy intellectual task – a World 2 attempt to grasp a World 3 object. ... This may involve a creative effort; the effort to grasp the abstract problem situation; if at all possible better than it was done before. Then he may produce his solution, his new theory. This may be put in linguistic form. ... Then he will critically discuss his theory; and he may greatly modify it as a result of the discussion. It is then published and discussed by others. ... And only after all these intensely intellectual efforts may somebody discover a possible far-reaching technical application, which acts on World 1.
> It is a fatal mistake to believe that there can be an adequate theory – psychological or behavioural, or sociological, or historical – of the behaviour of scientists which does not take full account of the World 3 status of science. This is an important point which many people are not aware of. These considerations seem to me decisive. They establish the objectivity of World 3, and its (partial) autonomy. And since the influence of scientific theories on World 1 is obvious, they establish the reality of the objects of World 3.

All of these arguments were unknown or unappreciated by the materialist philosophers who at Düsseldorf were emphatic in rejecting the World 3 of Popper. We shall return to this theme after considering World 3 in relation to Worlds 1 and 2 in the activities of the human brain.

7.2.3 Relationships Between the Brain and World 3

Figure 7-2 represents an attempt to give in outline the principal pathways for the flow of information for a single human person. The body has been separated from the rest of World 1 and is depicted as a small basal fragment with lines of communication in the afferent and the efferent pathways leading to and from the brain. In these respects Fig. 7-2 is a simplified version of Fig. 1-2, particularly with respect to the brain and the brain–mind liaison. The new feature is in the relationship to World 3. At the top of the diagram there is shown a fragment of World 3 encoded on a special component of World 1. Reference to Fig. 7-1 shows that such a component of World 1 is listed as 'material substrates of human creativity'. It could for example be a printed book, the paper and ink being in World 1 (the World 1 b category). The channels of communication shown by the vertical arrow to the left would be the optical pathway from the illuminated book to the eye for encoding and transmission to the brain (cf. Fig. 3-6). Reciprocally the vertical arrow to the right would be the line of communication when committing ideas to written expression in a book. Thus Fig. 7-2 illustrates very diagrammatically the ways in which we receive from World 3 objects which are coded on some World 1 base and in which we achieve expression of some idea (World 2) in some coded manner (World 3 a) on a material substrate (World 1 b).

It has seemed necessary to introduce one further complication into the brain of Fig. 7-2, and that is for memory stores which can properly be regarded as belonging to World 3, and which are shown as a brain compartment so labelled (World 3 b). The most striking exemplars would be the Homeric epics or the Icelandic sagas that were retained for centuries in the successive memory stores of bards before achieving an encoded expression in writing. But of course at an ordinary academic meeting much of the World 3 exchange is derived not from written manuscripts, but from the

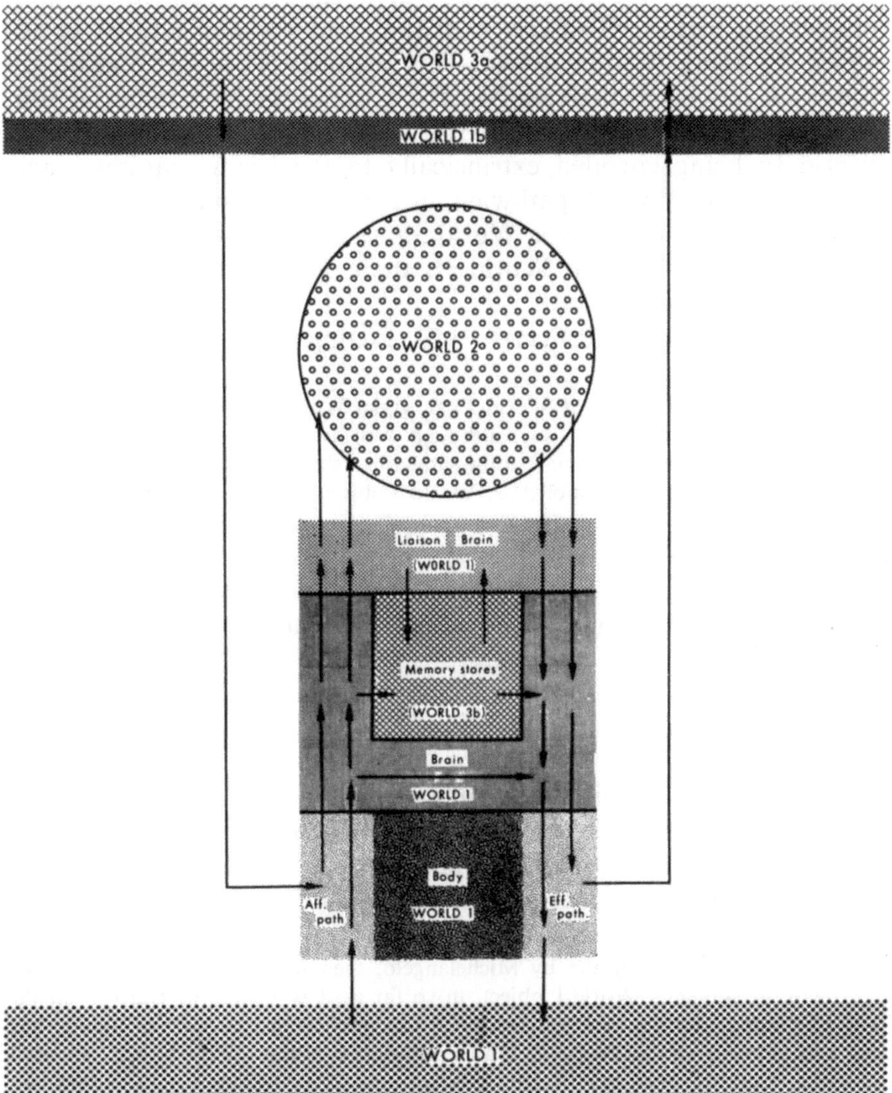

Fig. 7-2. Information-flow diagram representing modes of interaction between the three Worlds as shown by the pathways represented by *arrows*. It is to be noted that, except for the liaison between the brain and World 2, all of information occurs in the matter–energy system of World 1. For example, in the reading of a book, communication between the book and the receptor mechanisms of the eye is subserved by radiation in the band of visual wave lengths. It is to be understood that any individual can at will range widely in his relationship to World 3. For further description see text

memory stores of the participants, which can be objectified for communication just as effectively as if they had been objectified in advance by being written down. This memory store of World 3 in the brain has been distinguished from that coded on material substrates by being labelled World 3b, as against World 3a, in Fig. 7-2. World 3a being encoded extrinsically to the brain requires transmission along World 1 pathways in order to be subjectively experienced, i.e., the pathway is World 3a → World 1 → World 2, as depicted in Fig. 7-2. With World 3b there need not be any specific World 1 pathway.

Popper (Popper and Eccles, 1977, Dialogue XI) states:

> My thesis is not only that World 2 can grasp World 3 objects, but that it can do so directly; that is to say, although World 1 processes may be going on (in an epiphenomenal manner) at the same time, they do not constitute a physical or World 1 representation of those World 3 objects which we try to grasp.

He illustrates this with Euclid's proof that there exist infinitely many prime numbers. Euclid's proof was based on an intuitive grasp of the World 3 situation – of the infinite sequence of natural numbers.

7.2.4 Artistic Creativity and World 3

So far the World 3 discussion has been concentrated on scientific problems and their solution. It is important therefore to give examples of artistic creativity of World 3 objects. In Dialogue XI of the *Self and Its Brain* (Popper and Eccles, 1977) Popper states:

> If we look at a sculpture by Michelangelo, then what we see is, on the one hand, of course, a World 1 object, in so far as it is a piece of marble. On the other hand, even the material aspects of this, such as the hardness of the marble, may not be irrelevant to the World 2 appreciation of this World 3 object encoded in a World 1 substrate because it is the artist's struggling with the material, and the artist's overcoming of the difficulties of the material, which is part of the charm and the significance of the World 3 object. So I do not want generally to relegate the World 1 aspect of an encoded World 3 object to an epiphenomenon – but sometimes it is. If we have a book which is moderately well printed but not very well printed – not, say, a special edition – then the World 1 aspect of the book may be utterly irrelevant, and in a sense no more than an epiphenomenon, a sort of uninteresting appendix to the World 3 content of the book. However, both in the case of the Michelangelo statue and in the case of the book, what we – what our World 2, our conscious self – really gets

> in touch with is the World 3 object. In the case of the statue the World 1 aspect is important; but it is important only because of the World 3 achievement which consists in changing and modelling the World 1 object. In all cases, what we really look at and admire and understand is not so much the materialized World 3 object as the various World 3 aspects regardless of their materialization.

This argument of Popper is an effective answer to the criticism of World 3 by Passmore (1975), who refers to the work of a great architect, saying: 'Are we to call this a material substratum, this wonderful building, or an idea? There are ideas in the substratum, one can't use this language which contrasts substratum and idea'.

In the case of a musical composition the World 3 relationship is indeed very subtle and sophisticated. But is not this to be expected in respect of the wonders of musical creativity? In Dialogue III of Popper and Eccles (1977), Popper states:

> A musical composition has a very strange sort of existence. Certainly it at first exists encoded in the musician's head, but it will probably not even exist there as a totality, but, rather, as a sequence of efforts or attempts... ; so here there already arises a problem. Let us pose the problem in the following way. Clearly, Mozart's Jupiter Symphony is neither the score he wrote, which is only a kind of conventional and arbitrarily coded statement of the symphony; nor is it the sum total of the imagined acoustic experiences Mozart had while writing the symphony. Nor is it any of the performances. Nor is it all performances together, nor the class of all possible performances. This is seen from the fact that performances may be good or less good, but that no performance can really be described as ideal. In a way, the symphony is the thing which can be interpreted in performances — it is something which has the possibility of being interpreted in a performance. One may even say that the whole depth of this World 3 object cannot be captured by any single performance, but only by hearing it again and again, in different interpretations. In that sense the World 3 object is a real ideal object which exists, but exists nowhere, and whose existence is somehow the potentiality of its being reinterpreted by human minds. So it is first the work of a human mind or of human minds, the product of human minds; and secondly it is endowed with the potentiality of being recaptured, perhaps only partly, by human minds again. In a sense World 3 is a kind of Platonic world of ideas, a world which exists nowhere but which does have an existence and which does interact, especially, with human minds – on the basis, of course, of human activity.

The relationship of World 3 to books is more evident, but nevertheless has some subtlety, as is indicated by this imaginary conversation:

Holding a book in one's hand one asks:

'Have you read this book?' To which there is the strange reply: 'No, I haven't read *that* book but I have read a copy of this same book that I have at home.' This reply indicates a confusion of the material substrates (paper and ink) with the World 3 content. The question concerned World 3, the reply related to World 1. With translations we even consider the World 3 content of the book, and the usual reply to the above question would be: 'Yes, but in the German translation', for example.

7.2.5 Criticisms of the Objective Existence of World 3

Armstrong (1978) conceives World 3 as no more than:

> a noble metaphor based upon the relative independence of our cultural achievements (and failures) from this or that individual mind.

He admits that:

> a theory must be grasped more or less thoroughly if it is to have an effect upon a mind and so an indirect effect upon the world.

He goes on to discuss:

> two physicists with the same intellectual grasp of the same physical theory.... But granted that the encodings in different minds of the *same proposition* may differ, why should we think that each encoding reflects in its own way a World 3 object? The encodings will do all that is necessary causally. Each encoding of the *same proposition* will reflect the other ones. ... There seems no need to postulate in addition a World Three object standing over against all the encodings. (my italics)

In criticizing this statement Roger James (1978) makes the point that the phrase 'the same proposition' employed by Armstrong is precisely a World 3 object as defined by Popper. He goes on to say:

> It seems as though Armstrong is saying that, granted that we each speak English in our own slightly different way, but why pretend that there is such a thing as the English language?

This is an important statement in relation to the objectivity of World 3, because one of the most important constituents of World 3 is language in the descriptive and argumentative components. It can be considered as the greatest cultural achievement, there being firstly spoken languages and then some 5000 years ago conversion to symbolic representation was invented (cf. Lecture 6, Eccles, 1979a).

At Düsseldorf (World Congress of Philosophy, 1978), the materialist philosophers denied World 3 an objective status. For example it was argued by J.J.C. Smart that Chemistry was no more than chemists doing chemical things including utterances of their beliefs. The quotations from Popper (Section 11, Popper and Eccles, 1977) given above demonstrate the fallacy of this view of science, and it is argued at length by Popper in the form of an imaginary dialogue (Section 21, Popper and Eccles, 1977). By a strange irony these materialist philosophers have given Science a subjective status – beliefs and utterances derived therefrom. They are happy to do this because, on the identity theory of the brain–mind problem, all mental phenomena are 'identical' with materialist brain states. Admission of the objectivity of World 3 is anathema because it is non-material and cannot be explained away be some 'identity' relationship to material objects.

As against these materialist philosophers it can be stated that Chemistry is an immense World 3 object that has been created by the critical efforts of chemists. It is not fixed and eternal like a Platonic idea, but is always subject to remodelling in accord with critical evaluations and experimental testing. It is encoded in language, in chemical formulae, in chemical tables, in diagrams, in mathematical equations, and so has an objective existence (in World 3) independent of the whole concourse of chemists. If all chemists were wiped out by some nuclear catastrophe, but the World 3 records survived, the human survivors could eventually learn to recover Chemistry as a discipline from those World 3 records. This devastating thought experiment (cf. Popper, 1968a) is sufficient to establish the status of Chemistry as a World 3 object. And similarly for all other sciences. It should be remembered that Greek science was in the same manner recovered from the surviving World 3 documents.

So far we have accomplished merely an introduction to World 3. World 3 will be basic to the remainder of this lecture, and moreover the subsequent lectures will be built upon a World 3 philosophy. *It is important to recognize that World 3 is not genetically coded even to the slightest extent.* Genetic coding contributes importantly to the building of brains that have propensities for participating to varying degrees in the multifarious aspects of the World 3 of culture. But that participation has to be achieved as the result of a learning-memory effort. And so our theme leads on to the next section on remembrance.

7.3 Remembrance

7.3.1 Introduction

To a great extent our memories are concerned with World 3 objects from our earliest childhood, where the learning of a language was of paramount concern, and so on through the whole of our life. For example, I can recollect the long discipline in my academic life – that still goes on, as witness the intense effort in preparing these lectures. Not only is there the challenge for the remembrance of so much new material, but there is also the challenge to link these memories into coherent stories and also to strive for new insights given by creative imagination, which will be the final theme of this lecture.

It is usual to subdivide memories into motor memories and cognitive memories because to a considerable extent different neuronal mechanisms are concerned (cf. Eccles, 1979a, Lecture 9). Our present interest is entirely with cognitive memory, concerned as we are with the thoughts and ideas in the sciences and in the humanities. In my Gifford Lectures of last year, I devoted a whole lecture to learning and memory (Lecture 9, Eccles, 1979a). That lecture was particularly concerned with the structural and functional synaptic changes possibly related to memory, and with the evidence for the neural circuits concerned in the laying down of long-term memory. Amnesia resulting from clinical lesions gave valuable information to Kornhuber (1973) for building up circuits from the association cortex to the hippocampus and thence through the MD thalamus to the frontal association cortex. This basic circuitry of Fig. 7-3 has been developed into an hypothesis that attributed (Eccles, 1978, 1979a) learning to a specific synaptic potentiation which was brought about by a selective action on synapses in the cerebral cortex. It is in accord with the conjunction hypothesis developed by Marr (1969, 1970) in explanation of learning in the cerebellum and the cerebrum.

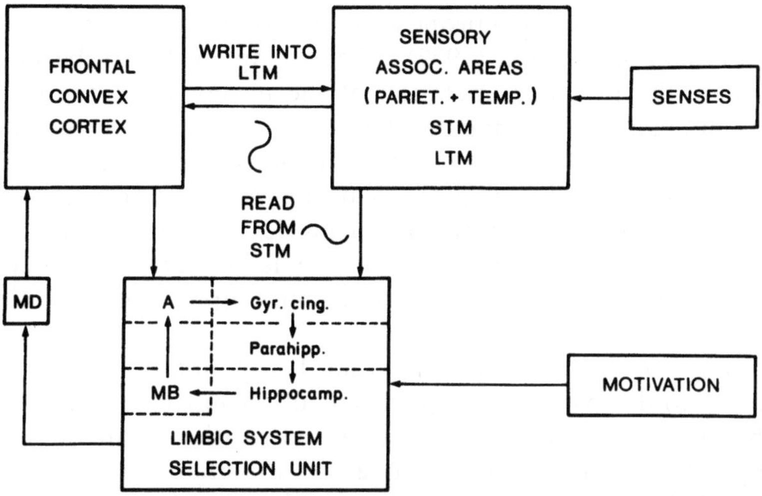

Fig. 7-3. Scheme of anatomical structures involved in selection of information between short-term memory (*STM*) and long-term memory (*LTM*). *MB*, mammillary body; *A*, anterior thalamic nucleus; *MD*, medio-dorsal thalamic nucleus. Kornhuber, H.H.: Neural control of input into long-term memory: limbic system and amnestic syndrome in man. In: Memory and transfer of information. Zippel, H.P. (ed.), pp. 1–22. New York: Plenum 1973

This selection theory of cortical learning is built upon known anatomical connectivities (cf. Fig. 7-3) and upon detailed synaptic structural arrangements in the hippocampus and in the neocortex and also upon the clinical lesions producing amnesia. But physiological evidence is most inadequate except for the hippocampal synapses in the postulated learning circuits, which do indeed exhibit prolonged potentiations (Bliss and Lømo, 1973; Misgeld et al., 1979) that correlate well with the synaptic hypertrophies seen electronmicroscopically (Fifková and van Harreveld, 1977), as has been described recently (Eccles, 1979b).

Presumably, because of the immense complexities of the possible circuits, there has been very little in the way of attempts to formulate theories concerning the manner in which neuronal structures participate in cognitive learning. My provisional ideas (Eccles, 1978; 1979a, 1979b) concentrated on the happenings during the laying down of memories and their storage, but the retrieval of memories was discussed only in general terms. There has not yet been a full appreciation of the modular operation of the neocortex (cf. Lecture 2) that

Szentágothai (1978a) had proposed on the basis of the radio-labelling by Goldman and Nauta (1977). The attempt will now be made to assimilate the modular operation of the neocortex to the selection theory of cognitive learning that has already been developed (Eccles, 1978, 1979a). On this basis it will be demonstrated that the synaptic modifications laid down in the learning process provide a simple and logical basis for retrieval of memories. It will appear that for the first time we can account for the laying down of memories and their retrieval in an hypothesis built upon the operation of the neural machinery of the brain.

It may be noted in passing that I will not be giving special attention to the experimental discipline of cognitive learning as studied by many psychologists because no correlation is attempted with the operation of the neural machinery of the brain, which essentially is regarded as a black box. However, this criticism does not hold for those schools of psychology that concentrate on the brain performance, such as is done by Milner, Weiskrantz, and Zangwill for example.

7.3.2 Role of Modules in Cerebral Learning

A necessary preliminary is to describe some special features of the cerebral cortex that qualify as essential structures concerned in the proposed learning process. In Fig. 2-6 the cortico-cortical afferent central to the module bifurcates in lamina I to form two horizontal fibres that run far beyond the picture, in fact for several millimetres. They are specially featured in Fig. 7-4 ascending in module C (ASS and COM) and running horizontally through modules B and A. The cells of origin of the ASS fibre would be a pyramidal cell of another module, often quite remote, while that of the COM fibre would be in the other hemisphere (cf. Fig. 2-7). By contrast the MA fibre in Fig. 7-4 is the axon of a Martinotti cell that lies deep in that same module, (cf. cell S6 in Fig. 8-7 of the preceding Gifford series, Eccles, 1979a).

These horizontal fibres form a dense meshwork in lamina I (Jones and Powell, 1970a), each fibre running for several millimetres. From their point of origin in the bifurcation, the horizontal fibres project in any direction in a plane parallel to the surface so that the overall result is a fairly uniform radiation, as depicted in Fig. 7-7 for the

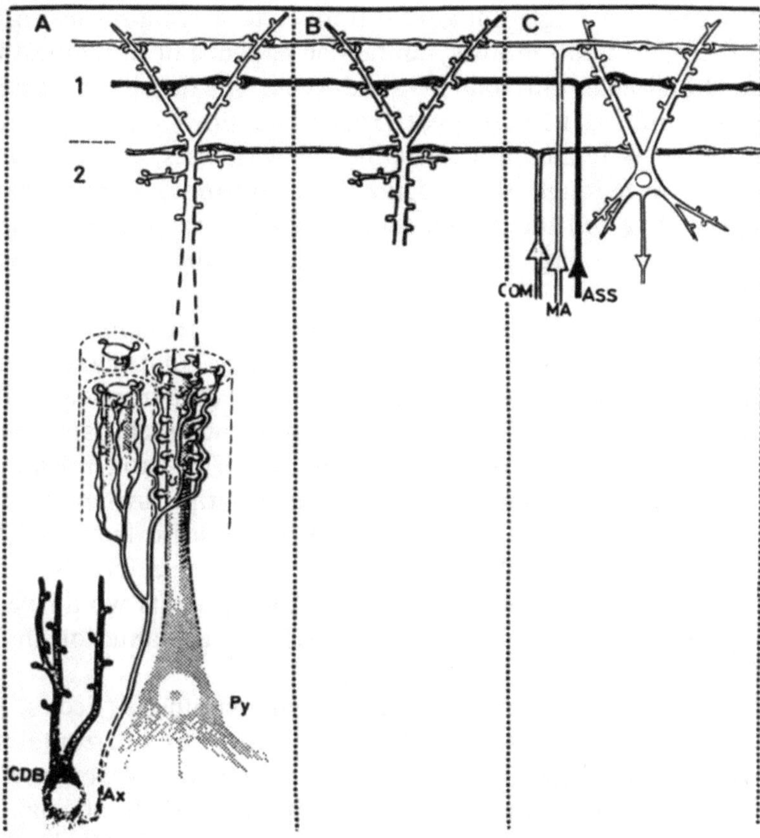

Fig. 7-4. Simplified diagram of connectivities in the neocortex that is constructed in order to show pathways and synapses in the proposed theory of cerebral learning (cf. Eccles, 1978). The diagram shows three modules A, B, C. In laminae I and II there are horizontal fibres arising as bifurcating axons of commissural (*COM*) and association (*ASS*) fibres and also of Martinotti axons (*MA*) from module C. The horizontal fibres make synapses with the apical dendrites of the stellate pyramidal cell in module C and of pyramidal cells in modules A and B. Deeper there is shown a spiny stellate cell (*CDB*) with axon (*Ax*) making cartridge synapses with the shafts of apical dendrites of pyramidal cells (*Py*)

horizontal fibres arising from the ascending axons in two modules. As shown in Fig. 7-4 these horizontal fibres make excitatory synapses on the spines of the bifurcating apical dendrites of pyramidal cells and stellate pyramidal cells. There are 1000 to 3000 of such spine synapses on the apical dendrites of a single pyramidal cell in laminae I and II (Szentágothai, 1978a), but very few synapses would be made on that cell by a single horizontal fibre (2 to 4 synapses in Fig. 7-4). Thus there is an enormous synaptic convergence of horizontal fibres – many hundreds – onto one pyramidal cell. The density of horizontal fibres in lamina I is very great (Jones and Powell, 1970a), but no quantitative estimates are available. Nevertheless, for our present purpose it is enough to recognize the enormous convergence on each pyramidal cell and the wide dispersal of horizontal fibres derived from axons surfacing in any one module, a radial spread of several millimetres (cf. Fig. 7-7).

The second important feature for the proposed hypothesis is the 'climbing fibre' arrangement, called a cartridge, that makes multiple excitatory synapses on the apical dendrites of pyramidal cells, as indicated in Fig. 7-4. Two cartridge arrangements can also be seen in Fig. 2-2 and in the top right of Fig. 2-6. The axon arising from cell CDB in Fig. 7-4 makes the complex multiple synapses on two adjacent pyramidal cells. The climbing fibres arise from cell SS in Fig. 2-6, which is directly excited by the thalamic afferents, labelled SPEC. AFF., but in association cortex the afferents would be from non-specific thalamic nuclei such as MD or the anterior thalamic (cf. Fig. 5-3).

On analogy with the cerebellar learning hypothesis (Eccles, 1977), it was postulated (Eccles, 1978) that synapses of horizontal fibres on pyramidal cell dendrites were hypertrophied if they were activated at about the same time as the cartridge activation of that same pyramidal cell, i.e., that there is a selection for hypertrophy on the criterion of approximate temporal conjunction. It is assumed that the climbing fibre synapses of the cartridge are homologues of the climbing fibre synapses on Purkyně cells, and that the horizontal fibres of laminae I and II are homologues of the cerebellar parallel fibres.

Thus in Fig. 7-4 the horizontal fibres formed by the axonal bifurcation of commissural (COM), association (ASS) and Martinotti cell (MA) fibres of one module (cf. Figs. 2-2 and 2-6) make synapses with the apical dendrites of the pyramidal cell (Py) of another module. If there is impulse discharge along those horizontal fibres at about the time of the cartridge synaptic activation, it is assumed that the synapses of those horizontal fibres would be selectively hypertrophied, as shown diagrammatically in Fig. 7-5, but not synapses made on that same apical dendrite by the COM, ASS and MA fibres coming from some module not then activated, e.g. COM and MA in Fig. 7-4. In the original formulation of the hypothesis of cerebral learning, the manner in which this selective hypertrophy could generate a learned cerebral performance was left undefined. This problem will now be considered in the light of the beautiful demonstration by Goldman and Nauta (1977) of the discrete modular structure of the neocortex and the conceptual development of cerebral operation deriving therefrom (Szentágothai, 1978a; Eccles, 1978, 1979a; cf. Lecture 2 above).

Figure 7-5 illustrates a selective hypertrophy of the synapses that the ASS fibre coming to module C from some other module (cf.

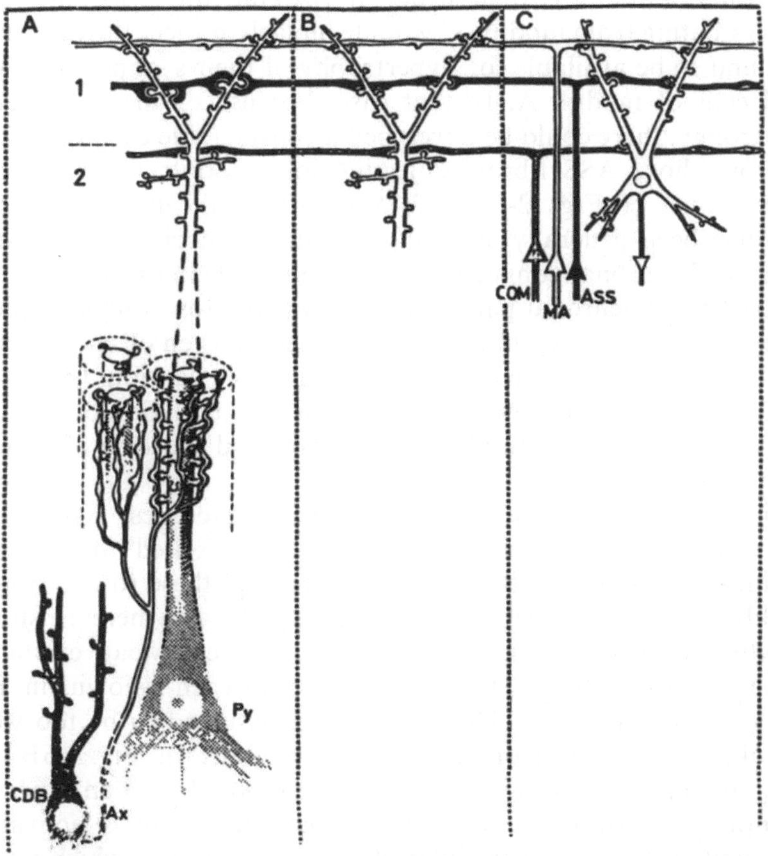

Fig. 7-5. Simplified diagram as in Fig. 7-4, but due to conjunction between the activation of the cartridge synapses of cell *CDB* in module A and the association fibre from module C there is a selective hypertrophy of those synapses on the apical dendrites of that pyramidal cell in module A, but not of the synapses of the *ASS* fibre on the pyramidal cell of module B

Figs. 2-5 and 2-7) makes on a pyramidal cell (Py) of module A. This effect arises as a consequence of conjunction with its activation by the cartridge synapse on that pyramidal cell. Because of the conjunction, these synapses are selected for hypertrophy, whereas there is no hypertrophy of the synapses by the ASS fibre on the pyramidal cell of module B, nor of the COM and MA synapses on either pyramidal cell. However it is implicit in the hypothesis that all these synapses potentially can be hypertrophied under the appropriate conditions of conjunction.

Provided that the distance between modules A and C is not too long, for example less than 1 mm, horizontal fibres of many ASS fibres radiating from module C would similarly project through module A and so be available for hypertrophy of their synapses on pyramidal cells of module A. In that case, because of the conjunction hypertrophy, there could be developed a considerable excitatory linkage from those ASS fibres projecting into module C (from some other module, cf. Figs. 2-5 and 2-7) to pyramidal cells of module A.

It has been generally assumed that the memories are stored in patterns of neuronal connectivity, the memory being retrieved when that particular neuronal pattern is reactivated. However this general concept lacks any diagrammatic representation, even of an elemental fragment, that could form the basis of a critical evaluation. In Fig. 7-6 there is an attempt to show how conjunction potentiation of horizontal fibre synapses could result in the storage and retrieval of a memory.

Figure 7-6A represents an assemblage of modules. Association fibre inputs (ASS) cause two modules to be excited to discharge ASS inputs into two other modules and each of these to two others. In addition between these two modular pathways there is shown a module in close proximity that would receive a weak excitation by the horizontal fibres (H) emanating from each (shown, in Fig. 7-4A for one ASS fibre). This excitation is assumed to be too weak to evoke a discharge from this central module. In Fig. 7-6B this background pattern of Fig. 7-6A has superimposed on it an experiential input that attracts interest and attention. There is activation of the limbic system, in particular of the hippocampus, which results in cartridge activation (Fig. 7-4; Eccles, 1978) of the central module (the curved arrows, C) as well as an association fibre input (ASS) to that module. But at the same time there is an increased activation of the ASS inputs into the background modules with conjunction

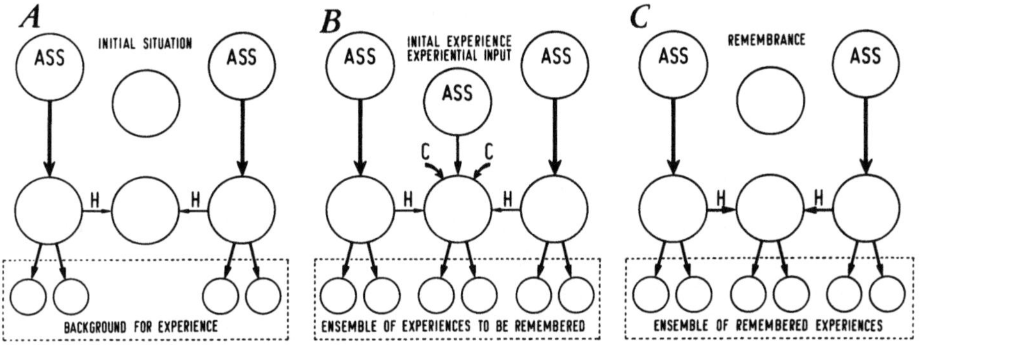

Fig. 7-6 A–C. An assemblage of modules as seen in plan from the cerebral surface in order to illustrate the proposed manner in which synaptic hypertrophy resulting from conjunction can cause the development of an effective modular activation. This could be the basis of a specific change in spatio-temporal modular performance, which is the coding of a learned memory. Full description in text

between the horizontal fibre input (H), to the central module and the cartridge input (C). As a consequence there is an enduring hypertrophy of these horizontal fibre synapses (thicker H in Fig. 7-6C) on the pyramidal cells of the central module. Because of the ASS and cartridge inputs into the central module, its pyramidal cells will discharge impulses that contribute to the ongoing pattern of modular discharge, as shown in Fig. 7-6B, which provides the neural basis of the experience that is labelled as 'ensemble of experiences to be remembered'.

Figure 7-6C shows the consequence of the hypertrophy of the H synapses on the central module. Now, in the absence of the experiential input of Fig. 7-6B, the background ASS inputs into the two modular systems will be able to excite the central module by virtue of the enduring enhancement of the synaptic potency of the H synapses, as is indicated by the stronger H arrows. So there will be the same ongoing modular pattern as in Fig. 7-6B. This pattern forms the neural basis of 'the ensemble of remembered experiences' in Fig. 7-6C.

It is important to recognize that Fig. 7-6 is a greatly simplified diagram of the neural events that are assumed to be necessary for a remembered experience. As seen in the previous lectures, there is much evidence that a conscious experience does not arise until there has been the development of a great spatio-temporal patterning of modular activity (cf. Popper and Eccles, 1977, Chaps. E2 and

E7), and necessarily it would be similar for a remembered experience. In Fig. 7-6C the ongoing modular pattern (the ensemble) is shown to be identical with that in Fig. 7-6B. That would correspond to perfect memory, and this would be the situation with the correct recovery of some simple task such as a number sequence or a name. But we are all familiar with imperfect recovery in even such a simple task, and with the repeated efforts we have to make to achieve a recall recognized as correct.

The neural basis of this situation can be identified in Fig. 7-6C. For example the adequate discharge of the central module is shown to be dependent on the convergence of potentiated H inputs from two adjacent modules that are activated sequentially by the background ASS inputs. If this activation is deficient, there will be deficiency in the modular activity that gives the 'ensemble of remembered experiences'. The ASS inputs in Fig. 7-6C may be triggered by a background experience that is provided by a specific sensory input as in Fig. 7-6A. Alternatively, we can recognize that an experience can be recalled in memory in the absence of such a background. In such a situation it is postulated that the background ASS inputs of Fig. 7-6C can be triggered by the action of the mind on the modular machinery of the brain, as has already been proposed in the dualist-interactionist hypothesis (Lectures 1, 2, 4 above; Popper and Eccles, 1977, Chap. E8; Eccles, 1978, 1979a, Chap. 10).

Brief reference should be made to the manner in which more complicated modular patterns can be developed from the principles derived from the elementary diagrammatic illustration in Fig. 7-6. In Fig. 7-7 the modules are shown in plan as seen from the surface, each being a circle of 250 μm in diameter. In this pattern the two background modules with ASS inputs are shown at about 500 μm apart with radiating horizontal fibres, and around both are modules with a gradually diminishing horizontal fibre activation, as shown by the progressive decrease in the density of H fibre distribution. Potentiation of the H fibre synapses to several modules as illustrated in Figs. 7-5A and 7-6C is shown by the convention of a thickening of some lines as they traverse these modules. It can be seen that around the two excited modules there will be a constellation of modules with an augmented input from H fibre synapses and that consequently are excited in the memory process to initiate a spatio-temporal pattern of ongoing modular activation as in the 'ensemble of remembered experiences' of Fig. 7-6C. Goldman and Nauta (1977)

Fig. 7-7. Diagram of a large assemblage of modules as seen in plan, each outlined by a circle. From two modules there are seen 38 radiating horizontal fibres that would travel far beyond this illustrated zone. Each is of course bifurcated in laminae I and II (cf. Fig. 7-4), so the number of radiating fibres is twice the numbers of fibres of origin, association, callosal and Martinotti (cf. Fig. 7-4). In several modules there has been hypertrophy of the synapses made by the traversing horizontal fibres as shown in Fig. 7-5, and this is shown by thickening of the lines. Further description in text

found that the radiotracer distribution was specially concentrated in laminae I and II (cf. Fig. 2-4 A) and spread well beyond the module defined by the 250 μm distribution in laminae III to VI. This distribution is attributable to the horizontal fibres. According to the present hypothesis, there is hypertrophy of the horizontal fibre synapses on some modules, so there should be a corresponding patchy distribution of the radioactivity in laminae I and II around a radioactive module. By careful quantitative study it should be possible to recognize if there is this predicted patchy distribution.

For the sake of simplicity the role of horizontal fibres arising by bifurcation of the COM (commissural) and MA (Martinotti cell axons) in Fig. 7-4 has been neglected. As already mentioned, the COM fibres are probably much less numerous than the ASS fibres, perhaps about 10%, but otherwise no special features need be introduced to explain their role in the memory process. Presumably they are adjuvant to the ASS fibres. Evidence for their effectiveness in the laying down of memory traces is given by the finding that after unilateral hippocampectomy such traces can still be formed, but in a deteriorated manner, for special hemispheral functions on that side (Milner, 1974). Presumably the commissural fibres carry these functions to the other hemisphere for conjunction with the hippocampal input to that hemisphere.

The contribution of the horizontal fibres from Martinotti cells is probably very important in memory. For example Jones and Powell (1970a) suggest that horizontal fibres are mostly formed from Martinotti axons. It can be assumed that Martinotti cells are to a considerable extent activated synaptically by the input of association and commissural fibres to that module; but the excitatory input from the thalamic nuclei also relays in Lamina IV to interneurones such as spiny stellate and neurogliform cells whose axons descend to excite the Martinotti cells in lamina 6 (Szentágothai, 1978a, Figs. 10, 13). Such connections give additional opportunity for thalamic inputs to contribute to the horizontal fibre activation that is concerned in the laying down of memories.

In the recently proposed theory of cerebral learning (cf. Eccles, 1978) attention has been concentrated on the horizontal fibres. No reference has been made to other excitatory synapses on pyramidal cells such as are formed by spiny stellate cells that are activated by inputs both of the ASS–COM afferents and of the specific thalamic inputs. If there were conjunction between activation of the cartridge synapses and the synapses that these cells make on the pyramidal cells, there should also be hypertrophy of these synapses. Presumably the synaptic potentiation so produced could contribute to the memory trace. Emphasis has been placed on the horizontal fibres for two reasons. Firstly there is the analogy with the parallel fibres in cerebellar learning. Secondly the horizontal fibre system gives special opportunity for the incorporation of new modular systems into the spatio-temporal patterns that are the neural counterpart of the memory.

It is convenient to describe memories as being encoded in 'data banks' in the neocortex. How can these be defined in relation to the present modular hypothesis? By a data bank we can understand the whole ensemble of modules that by potentiated synaptic connectivities tend to be activated in some specific spatio-temporal manner and that thus encode the experience to be remembered. We must recognize that it is the spatio-temporal pattern that encodes the data bank, not any particular module. Any one module can participate in an indefinitely large number of different spatio-temporal patterns (cf. Fig. 2-9). So the potentiality for storage of memories as data banks is beyond estimate even though there are no more than 4 million modules for encoding the patterns (Eccles, 1978, 1979a).

The basic structure of modular connectivities by association and commissural fibres is laid down in the ontological development of the neocortex. In the primate it is already present a few days after birth (Goldman and Nauta, 1977). It would be expected that it would remain unchanged throughout life because there is no developmental process for establishing new tracts. These modular connectivities provide, as it were, the skeletal structure of the neocortex. Plastic changes in the cerebral performance presumably are brought about as a result of changes in the existing connectivities between modules. These changes can be effected by the increased synaptic effectiveness of the horizontal fibres as proposed in this hypothesis of cerebral learning (Fig. 7-5; Eccles, 1978). As a consequence, as indicated in Figs. 7-6C and 7-7, there are opened up pathways that hitherto were ineffective. We might term the situation depicted in Figs. 7-6C and 7-7 as 'modular jumping'. Figure 7-7 illustrates the patterned distribution of this new development of modular connectivities. It may be regarded as a very simplified model of the cortical changes responsible for cognitive memory. Hitherto there have been no models for cognitive memory that bear any relationship to the anatomical structure of the neocortex.

7.3.3 Conclusions

The attractiveness of this modular hypothesis of cognitive memory lies in the fact that it utilizes the modular activity for storage and retrieval in the whole neocortex. Our cognitive memories do cover the whole cerebral performance. No special operation is required for retrieval. This is accomplished by the modular operation with its patterned remembrance encoded as indicated in Fig. 7-7.

Figures 7-6 and 7-7 are complementary in that both display in a different convention the convergence of ASS inputs projecting from two different modules, which are not of course the modules of origin of the horizontal fibres. Because of the synaptic hypertrophy these convergent paths may be able to activate that recipient module sufficiently to cause it to generate a discharge of impulses and so to add its contribution to the spatio-temporal pattern of modular operation. Thus a new component is added to the pattern, and as seen in Fig. 7-6 this component resembles that initially occurring in the laying down of the memory trace. Of course this is a greatly over-simplified account. We can conjecture that any one cognitive memory would have a neural counterpart in the patterned performance of hundreds or even thousands of modules that owe their performance to the plastic changes (cf. Figs. 7-5, 7-7) induced in the learning process and that their replay participates in the recovery of the memory. Memories can be recovered by means of the ongoing sensory input or by the mental process of deliberate recall by the operation of the mind on the brain, as indicated in Fig. 1-7.

7.4 Creativity

7.4.1 Creative Imagination in Science

In our present context creativity is a distinctively human activity. It is human creativity that has given us the whole of World 3. Every aspect of culture and civilization can be regarded as the fruits of human creativity. The scope is immense, but here our special task is to relate the creative process to the hypotheses that we have been developing on the brain–mind problem. Have these hypotheses any explanatory power in respect of creativity? We must expect no more than to display this in a general manner especially in relation to scientific creativity.

Medawar (1969) speaks of the thinking up of a hypothesis as:

> the generative act in scientific discovery. ... 'Creativity' is a vague word, but it is in just such a context as this that we should choose to use it. ... That

'creativity' is beyond analysis is a romantic illusion we must now outgrow.... We can put ourselves in the way of having ideas, by reading and discussion and by acquiring the habit of reflection, guided by the familiar principle that we are not likely to find answers to questions not yet formulated in the mind.

Monod (1975) asks the question:

whether you would agree that creativity in science ... is a very specific kind of creativity, which cannot be done without a very clear concept of a certain number of basic postulates that have to be respected, otherwise you are not creating scientific knowledge ... we must recognize the extreme specificity of this type of creativity.

Medawar's statement leads on to the question of creative imagination. Only with a highly developed imagination can we create ideas that form the lead into any worthwhile scientific investigation. Popper (1975) expresses this well:

I think that it is possible to subsume both the production of new ideas and subjective simulation under one category – that of bold imagination.... And imagination, it should be stressed, is needed not only for thinking up new ideas, but also for criticizing them. You need imagination especially in order to imagine new situations which might be unfavourable to your ideas, and also in order to apply your ideas to these new situations.

'Subjective simulation' had just been defined by Monod as the process whereby we simulate subjectively the events around us. The phenomenon itself and the sources involved are internally represented. For example a physicist would identify himself with an electron or particle and ask what would I do if I were that particle? This is an excellent example of the imaginative process!

Bronowski (1973) makes this point very well when describing Niels Bohr's remarks to Heisenberg:

When it comes to atoms, language can be used only as in poetry. The poet, too, is not nearly as concerned with describing facts as with creating images

and Bronowski comments:

That is an unexpected thought; when it comes to atoms, language is not describing facts but creating images. But it is so. What lies below the visible world is always imaginary, in the literal sense; a play of images. There is no other way to talk about the invisible in nature, in art, or in science.

And with subatomic particles even the names are poetic – strangeness, colour, charm.

The experience of one image is evocative of other images, and these of still more. When these images are of beauty and subtlety, blending in harmony, and are expressed in some language – verbal, musical or pictorial – to evoke transcendent experiences in others, we have artistic creation. Creative imagination is the most profound of human activities. It provides the illumination that gives a new insight or understanding. In science that illumination leads to the creation of a new hypothesis which embraces and transcends the older hypotheses. Such a creation of the imagination has immediate aesthetic appeal in its simplicity and scope; it must nevertheless be subjected to rigorous criticism and experimental testing. There is a whole literature on the romantic experiences of sudden flashes giving a new imaginative insight. It comes to us rarely and unfortunately is usually falsified in the cold light of reasoned criticism and by experimental testing. But these experiences do add to the zest of the scientific life.

Popper (1975) gives excellent advice:

> The important thing is that you should *care* for your problem: that you should enter into your problem situation in such a way that you almost become part of it ... to get new *and* good ideas, what one must do is first to produce new ideas, and secondly, to criticize them. This is also the way in which one gets a grip on a problem. ... If you produce many different solutions to the problem and they all do not work, not only will you have discovered that, ... but you will also have become a sort of expert in the problem, and will be in a good position critically to evaluate new suggestions, whether suggestions of your own or of other people. It is in this way that you become acquainted with a problem and the problem becomes part of you.

Later, when referring to incompatible beliefs, Popper suggests that the resulting feeling of discomfort is the thing that led to Greek rationality, and so it marks the beginning of science.

7.4.2 Creative Imagination and Brain–Mind Problem

We now face the challenge of somehow relating creative imagination to the respective roles of brain and mind, particularly in the light of the modular theory of neocortical action. On the modular theory we have to conjecture that creativity is dependent on the development of novel spatio-temporal patterns in the neural machinery. How can that be? If we become expert in a subject and immersed in it, as described above by Popper, it would be expected that the modular counterpart would be interweaving of spatio-temporal patterns which would interplay in the most subtle and rich variety. This would correspond to the thoughts that press upon us when struggling conceptually on a problem. The patterned forms would be dependent in part on the learning that gives new potentialities for pattern formation as indicated in Figs. 7-6 and 7-7. Meanwhile, the conscious mind struggling with the problem can modify the modular patterns as described in Lecture 4 and experience these changes as described in Lectures 1 and 2. Then a unique spatio-temporal pattern may evolve out of all this patterned modular activity and be consciously recognized as a new and valuable idea. It will be held in memory or expressed symbolically in language if it becomes clear enough for such formulation. Then the conscious mind draws on memories to criticize this new idea and maybe to modify it through its modular counterpart. An important additional point is that the intense spatio-temporal modular activity can be continuing when conscious attention is directed elsewhere. Thus we can account for the subconscious creation of at least a partial solution that can be completed by conscious critical thought.

Monod (1971) when speaking of scientific imagination writes:

> This mental reflection, at the deeper level, is not verbal; to be absorbed in thought is to be embarked on an *imagined experience*, an experience simulated with the aid of forms, of forces, of interactions which together only barely compose an 'image' in the visual sense of the term.

Here surely Monod is talking in World 2 terms of imaginings that are located in the inner sense of World 2 (cf. Fig. 1-7) and that may eventually be more clearly defined in linguistic or mathematical symbols as the liaison brain is effectively brought into action. Imagination has strong emotional overtones so we can conjecture that the associated neocortical spatio-temporal patterns have powerful reciprocal connections to the limbic system and the hypothalamus (cf. Lecture 5).

In order to participate at a high level, the brain of the subject must be richly endowed with resources for modular patterning that can be consciously controlled. There must be a high ability to store and retrieve remembrances over a lifetime. Our ignorance of higher cerebral functions is all too evident when speculating on the neural basis of imagination for example. Yet it is this property of the brain that is of paramount importance. Imagination cannot be learnt. It is an endowment that we can be overwhelmingly grateful for – that is my personal attitude to my own endowment which has enabled me to do whatever I have done.

The mental aspect of scientific creativity has already been dealt with in the preceding section. Here we can refer again to the continual driving interest never to let the problem rest. Moreover there is the wonderful gift of imagination that we may suppose to be derived in part from special neural design, but which gives thoughts transending the neural performance together with an ability to identify oneself with components of the problem, as in the identification with an electron, as suggested above by Monod. Intensely experienced images have an amazing vitality so that the subject becomes obsessed with them. Another mental attribute is the courage that is needed in daring to go beyond the conventional and the dogmas of the establishment. But it must be tempered by the humility required in the exercising of critical restraint on a too ebulient creativity. And so we enter on the field of values that will be the theme of Lecture 9.

Lecture 8

Altruism, Pseudaltruism and Aggression

Résumé

It is pointed out that much confusion has been generated by sociobiologists who denote as altruism those aspects of animal behaviour where an individual acts in some sacrificial role and in a manner that 'appears' to suggest concern for others of the same species. With the social insects this behaviour is wholly instinctive and hence dependent on genetic inheritance. Even with higher animals such behaviour is largely instinctive, though in part it may be learned. Only with human behaviour do we come to a quite different behaviour that when fully developed is a moral act freely chosen on the basis of *concern* for another's welfare and with *intention* of conferring benefit.

In order to sharpen the argument there is a brief introduction to genetic inheritance and to concepts of genotype and phenotype.

A consideration of the sociobiological concept of *kin-selection* leads on to the formulation of a clear distinction between instinctive sacrificial behaviour and true altruism. The suggested terminology is *pseudaltruism* for the former, the term *altruism* being reserved for its customary usage of being a moral act freely chosen. As defined in this way altruism is a creative and learned response and is not as such genetically inherited. What is inherited is the human brain constructed with the propensity to learn altruism. It is pointed out that there is an exact analogy with human language where genetic coding builds a human brain with propensity for language, but its usage for a language depends on learning that language.

On indirect evidence it is conjectured that altruism can probably be traced back to the postulated food-sharing of Australopithicines, but the earliest fossil evidence came very late in hominid evolution, the first known example of compassionate behaviour being in the Shanidar cave 60,000 years ago.

The *words altruism and pseudaltruism are thus applied to two quite different kinds of behaviours.* There is strong criticism of Wilson's terminology of hard-core and soft-core altruism which roughly parallel pseudaltruism and altruism, but which lead to most destructive and cynical comments on human society.

In contrast, in his Gifford Lectures, Sherrington referred to the mysterious origin out of the evolutionary process of morality and thus of altruism.

As we turn to the opposite kinds of behaviour that go under the names of aggression and violence we find more in common between animal and human behaviour. There has been general acceptance of the 'vogue' idea that man is the most aggressive of all higher animals against his own kind. This statement is based on flimsy data interpreted by anti-human prejudice, and not on the scientific study of animal behaviour. It now has been shown quite surprisingly that we are amongst the most pacific of the higher animals.

Nevertheless there is great danger in the growth of human aggression fostered as it is by its continual exhibition in the media. Always our society is threatened by the aggression of *homo praedatorius*, who even outdoes animal aggressors. Sociobiology could provide valuable insights into this most traumatic field of human aggression.

8.1 Introduction

As we pursue our enquiry into the human psyche, we come to the very controversial topics of altruism and aggression.

In his monumental treatise on sociobiology, Wilson (1975) has created a new science of animal behaviour that is based principally upon biological evolution and the genetic explanation of heredity. Sociobiologists (Wilson, 1975, 1978 and Dawkins, 1976) make great display of their strictly scientific methods. It is the claim of Wilson that the methodology and findings of sociobiology are applicable also to the social behaviour of *Homo sapiens*, and that his new synthesis would provide for anthropology and sociology an objective scientific basis. The strong criticisms to which Wilson was subjected were based on his premise that human behaviour is genetically determined, as may be seen in the critical discussion with Marvin Harris (Wilson and Harris, 1978).

In this lecture, I will be maintaining that sociobiologists have misunderstood human nature because of their concentration on models of animal behaviour largely derived from studies on social insects. Such behavioural terms as altruism and selfishness have been used indiscriminately for insects and for human beings. The use of the word 'altruism' by sociobiologists has been at the root of the misunderstandings. As Stent (1977) has pointed out there are two quite distinct meanings, and Stent uses the words *altruism*$_1$ and *altruism*$_2$ in his critical discussion of Dawkins' (1976) book. Altruism$_1$ is the customary use of the word, altruism, in the discipline of the ethics of human behaviour. Altruism$_2$ is applied to the self-sacrificing behaviour of animals, particularly of the social insects. As a convenient modification of this dual terminology, I will suggest pseudaltruism for altruism$_2$, keeping *altruism*$_1$ for its traditional usage in respect of human actions. The prefix 'pseud' is deserved because sociobiologists have disfigured the word 'altruism', which had been entirely human in its connotation, by applying it to some superficially similar, but fundamentally different animal behaviours.

The importance of this distinction can be illustrated by reference to human language. All would agree that what human beings inherit is a highly specific ability of their brains to communicate by language (cf. Popper and Eccles, 1977, Chaps. P3 and E4; Eccles, 1979a, Lecture 7). But there is no propensity of any brain to initiate speaking in one language or another. The brains of all races of people are equipotent in respect of the languages of the world. The language learnt is the language heard. Languages are the most important components of culture (World 3) and their development and maintenance are due to cultural evolution. I am referring specifically to human language, not to the so-called animal languages that are merely expressions and signals (cf. Popper and Eccles, 1977, Chap. P3, Sect. 17). Without linguistic communication in some form a baby cannot develop into a human person, as has been argued in my previous Gifford series (Eccles, 1979a, Lecture 7).

The clear separation between the roles of biological evolution and cultural evolution in respect of language gives the guide lines for making a similar distinction for altruism. What I am calling *pseudaltruism* would correspond to all behaviour patterns that have been developed in biological evolution. With human beings they remain as the vestigial remnants of our animal ancestry. We have such vestiges in the simpler forms of linguistic communication, such as calls and cries. I will later argue that much of aggression is another such vestigial remnant. On the other hand *altruism* develops in our cultural evolution, just as does language. The propensity to altruism is inherited as a property of the neural machinery of the human brain, and this propensity arose in biological evolution. The practice of altruism is the *intention* to act for the good of others. It is one of the glories of our human culture and must be learned just as we learn a language. It is no more inherited than is the speaking of a particular language.

The emotionally charged conflicts over sociobiology arose because there is a dogmatic assumption that the study of animal behaviour gives the key to the understanding of human behaviour in every aspect. My criticism has the aim not of discrediting sociobiology, but of showing how its claims have been presented as if it were a well established scientific discipline, whereas there are tremendous unknowns integral in its naive assumptions.

Sociobiology is a challenging discipline with the most fascinating stories of animal behaviour. The stories that Wilson (1975) assembles in his great treatise *Sociobiology, a Modern Synthesis*, are better than any science fiction. We are overwhelmed by the scholarship and the excellence of the descriptions and illustrations. However I share with Stent (1978), for example, reservations about the way in which these observations on animals are used to support dogmatic statements in the field of human psychology and ethics. After the controversial reception of his first book Wilson (1978) has followed by a book more specifically directed to the human situation. This book is a sequel to the last chapter of Sociobiology, and we are told on the jacket that:

> On Human Nature begins a new phase in the most important intellectual controversy of this generation. Is human behaviour controlled by the species' biological heritage? Does this heritage limit human destiny?

Sociobiologists are not remarkable for their modesty. This can also be said of another recent hybrid discipline called psychobiology (Uttal, 1978). There seems to be something in hybrid vigour!

8.2 Genetic Inheritance

We can assume that all animal behaviours have developed under the process of biological evolution. However, sociobiologists have been too uncritical and too optimistic in forcing explanations of altruistic and selfish behaviour patterns on the basis of genetic quantification. It is important therefore to clarify the use of the word 'gene' by a very simple account given for the purpose of this lecture. For an authoritative account, reference should be made to a text on biology such as *Biology* by Villee (1972) or *Biology Today* (1975).

Ever since the discovery by Beadle and Tatum the word gene has been restricted to that segment of the double helix of nucleotide sequences of DNA which is responsible for the synthesis of a particular enzyme or other protein. In the transcription–translation process a sequence of three nucleotides is responsible for a particular amino acid. This triple coding in serial array is read off from the nucleotide strand of the DNA for amino acid after amino acid of the protein being synthesized. Since there are hundreds of amino acids in serial array in a protein, a gene would be a segment of DNA involving a thousand or more nucleotide sequences. There are at least 10,000 different kinds of protein in the human body. The number of genes in each cell is estimated to be at least several hundred thousand, though at any one time the great majority are repressed. Moreover only a fraction of the 3.5 billion nucleotide sequences in human DNA has a genetic function, the remainder having no known function.

In all body cells there are pairs of similar chromosomes (23 pairs in humans) and each pair is constituted by one maternal and one paternal. In the formation of the sex cells there is a halving of the number of chromosomes. This reduction is not accomplished by a simple separation of each chromosome pair into the maternal and paternal components. There are crossing-overs and recombinations in quite indeterminate ways so that any one sex cell has its 23 chromosomes with mixed maternal and paternal inheritance, as is indicated in Fig. 8-1. Thus in the coming together of the sex cells in the process of fertilization, there are immense opportunities for genetic diversity, so much so that any one individual has a genetic constitution that is unique for all time. The only exceptions are identical twins, where one fertilized ovum makes two individuals with identical genetic inheritance.

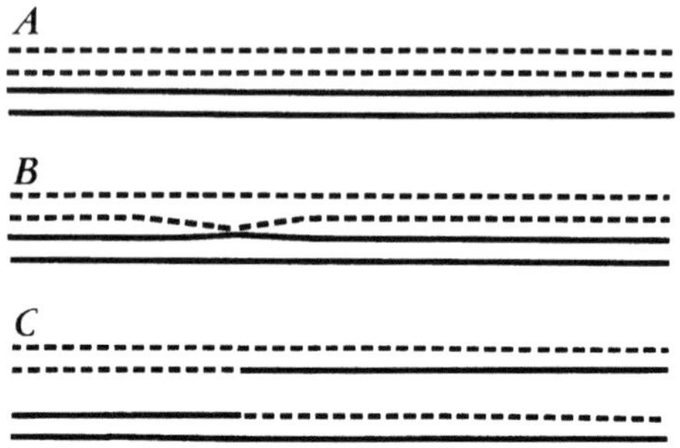

Fig. 8-1 A–C. Illustration of the process of crossing over and genetic recombination. In **A** two homologous chromosomes line up, each composed of two chromatids. In **B** the adjacent chromatids, one from each chromosome, are about to meet to form a chiasma. These chromatids break and in the crossing over and in **C** recombine to create two new gene arrangements

There are simple examples of a unitary correspondence between a single gene and an inherited disability, as for example in phenylketonuria, where a single pair of recessive genes results in the failure to synthesize an enzyme that metabolizes phenylalanine, one of the constituent amino acids of proteins. As a consequence there is a defective brain development. Another example is sickle-cell anaemia which is due to a defect of a single nucleotide pair of the gene for manufacturing the protein, haemoglobin, which consequently has an abnormal amino acid sequence and abnormal properties. However such instances are rare. Usually many genes co-operate in the development of some particular attribute of the phenotype.

The aim of the quantitative study of inheritance by sociobiologists has been to give scientific explanations of the way in which natural selection works in social animals. However, there are tremendous unknowns in the phenotypic development from the genetic instructions to the animal performance in some instinctive manner. For example, with a complex instinctive behaviour pattern such as that involved in caring for the young by birds and mammals, there is no scientific understanding of how the genetic inheritance (the genotype) controls the building of the neural machinery responsible for the behaviour.

Yet it is with respect to such complex behaviours that sociobiologists talk about genes for altruism or selfishness or spite. In fact, *The Selfish Gene* (Dawkins, 1976) is the 'catchy title' of a book, the gene being defined by Dawkins in a glib way so that he can use it for spinning his clever webs of argument and calculation. And of course only by the wildest metaphor can these genes be personified to exhibit behaviours as if they were selfish. Admittedly Dawkins does not claim that his book is for the expert, but he does claim that he is not writing science fiction, though he gets close to that on many occasions. For example, when speaking of the history of genes from the origin of life we are told on the jacket:

> They are in all of us; they created us, body and mind; and their preservation is the sole reason for our existence. ... Thus Richard Dawkins introduces us to ourselves as we really are – throwaway survival-machines for our immortal genes. Man is a gene machine: a robot vehicle, blindly programmed to preserve his selfish genes.

Stent's (1978) criticism is relevant:

> This presently utterly obscure relation between the animal and its genes is one of the hidden skeletons in the cupboard of neo-Darwinian evolutionary theory. Since it is clear that natural selection selects the overt features of the animal, but since it is also clear that what is passed on to future generations are the covert genes, we must know the ontogenetic relation between genome and phenome before we can understand how evolution really works. To cover up this conceptual deficit sociobiologists arbitrarily define the gene as that hereditary unit which natural selection happens to select. But since that unit is certainly not what geneticists mean by 'gene', the sociobiological 'Selfish Gene' is neither selfish in the context of morality, nor is it a gene, in the context of genetics.

I will end this section by a general criticism of sociobiologists that has recently been made by Campbell (1975), and with which I agree:

> Without giving a single personality test or doing any systematic behavioural observations, they (the sociobiologists) are willing to talk about characteristics of 'species personality' using terms like *altruism, coyness, spite, jealousy, selfishness, deceitfulness, greediness and co-operativeness,* with the explicit assumption that there are specific genes determining these traits. ... In contrast with psychology's experimental behaviour genetics ... these population geneticists assume that genes exist for every behavioural tendency. Aware that there are millions of genes, each with multiple effects and interactions, population geneticists assume that an unspecified subset is available to influence any behavioural trait in any direction. Then, for simplicity's sake, they plot the hypothetical fate in a population for a single gene, for example, one determining self-sacrificial altruism in the form of bravery in group defense.

This critical appraisal by a distinguished psychologist–philosopher is a good introduction of the topic of kin selection.

8.3 Kin Selection

The ideology underlying the development of sociobiology relates to the Darwinian concept of survival of the fittest. Two questions arise:

What is meant by survival?

What is meant by fittest?

When considering these questions in the context of animal societies, the fittest in the Darwinian sense is not of an individual, but of the population. Wilson (1975) expounded the 'kin selection' theory, which derives from the fact that the genes of an individual are to some extent duplicated amongst close relatives – halves to brothers and children, quarters to grandchildren, eighths to first cousins, and so on. And *fitness is measured by the reproductive success of the ensemble.*

Hence the sacrificial death of an individual can enhance the survival of its genes because of the similar genes in the kin that survive by virtue of the sacrifice. A good example is the alarm call of birds. There are many examples in the social insects such as the sacrificial stinging of an intruder by a honey-bee, or the sacrifice of a soldier ant. These sacrificial behaviours can thus be recognized as increasing the fitness of the social group; hence their genetic coding becomes intelligible on Darwinian terms. The sacrificial individuals act instinctively, and so their behaviour qualifies as a pseudaltruism, as will be discussed in the next section.

According to Dawkins (1976) an entity is altruistic if it behaves in such a way as to increase another entity's welfare at the expense of its own, *welfare being defined as chances of survival.* A good example is the alarm call of birds. Even though the calling bird may be sacrificed to the predator, it may have increased the chances of survival of its kin in the flock, and these would carry replicas of some of its genes.

Following Hamilton, Dawkins (1976) presents the mathematics of gene survival, in an entertaining chapter entitled 'Genesmanship' on the theme that the preservation of genes is the sole reason for our existence! On the basis of the gene fractions given above, it is proposed that animals care more for the welfare of their nearer kin than for more distant, and not at all for strangers, which strangely should include wives! For example you should be more prepared to undergo grave risks in saving the life of your brother ($^1/_2$) than for your first cousin ($^1/_8$) and hardly at all for third cousins ($^1/_{128}$). It is all worked out mathematically on a cost-benefit basis for genes!

Wilson (1975) gives three definitions in relation to kin selection together with an amusing illustrative diagram (Fig. 8-2) that seems to be humanly orientated, but it really obtains for animal behaviour in general:

> When a person (or animal) increases the fitness of another at the expense of his own fitness, he can be said to have performed an act of *altruism*. ... In contrast, a person who raises his own fitness by lowering that of others is engaged in *selfishness*. ... Finally, a person who gains nothing or even reduces his own fitness in order to diminish that of another has committed an act of *spite*.

In Fig. 8-2 the kin has been reduced to an individual and his brother, who shares half his genes by virtue of common descent, as shown by the blackened areas of the body. These areas represent what is known as 'genetic fitness'. The symbols of vessel and axe are obvious!

We are reminded of Campbell's (1975) criticism at the end of the genetic section. It is important now to disentangle the purely instinctive behaviour patterns of animals from the patterns of human behaviour in which instinctive behaviour is subjected to cultural control. The next section will be oriented to instinctive behaviours superficially resembling altruism and which I have proposed naming 'pseudaltruism'.

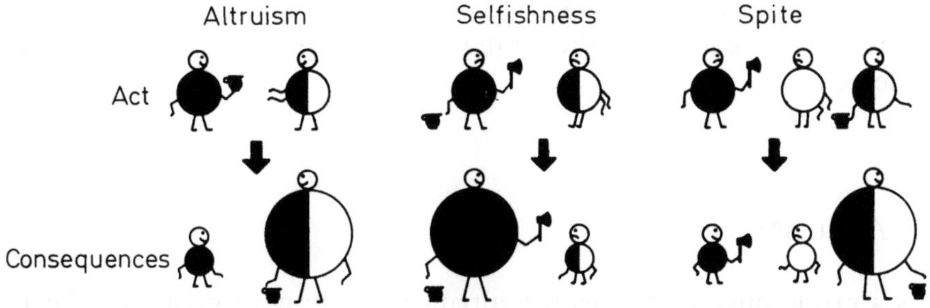

Fig. 8-2. The basic conditions required for the evolution of altruism, selfishness and spite, by means of kin selection. The family has been reduced to an individual and his brother; the fraction of genes in the brother shared by common descent is indicated by the shaded half of the body. Genetic fitness is indicated by the size of the black areas. Further description in text. Wilson, E.O.: Sociobiology: The new synthesis, pp. 697. Cambridge, Mass.: Belknap, Harvard University Press 1975

8.4 Pseudaltruism

It can be anticipated that eventually biological evolution with its offspring, sociobiology, will be able to account for many diverse behaviours that have gone under the name of altruism. However, as Wilson (1975) recognizes, this achievement will depend on immense progress in neurobiology. There is a formidable gulf to be bridged between the genetic constitution of the fertilized ovum and the fully fashioned behaving animal.

Pseudaltruistic behaviour of animals appears always to be instinctive, but is usually not sacrificial as in the cases quoted above of honey-bees, soldier ants and warning birds. For example, birds feed their young at immense effort, thereby ensuring that their genetic endowment will be carried on into the next generation. There are at present several tragic examples where, as birds become extinct, this genetic inheritance fails, the genes of that species being lost for all time.

There is similar pseudaltruistic behaviour in the suckling of young by mammalian mothers. The instinctive character of the act is indicated by the failure to safeguard the feeding of the young after weaning. For example, Bertram (1975) gives a moving account of the sorry plight of lion cubs after weaning, some dying of malnutrition because of neglect. It is plausible to suggest that the suckling of babies by human mothers is also an instinctive pseudaltruistic behaviour that is dependent on biological evolution. In an unsophisticated society suckling is as natural as eating when you are hungry. This instinctive human act is meshed in with a truly altruistic behaviour that continues with the caring for the young after weaning and right through adolescence.

8.5 Altruism

It is a crude mistake to identify altruism solely with great sacrificial acts such as diving into a treacherous sea to rescue a drowning person at great risk. That kind of altruism would have special appeal to sociobiologists whose scientific orientation is to the social insects! However, I do not wish to decry or undervalue heroic altruism. In my adopted country there is a dedicated veneration for their great hero, Arnold von Winkelried, who, at the battle of Sempach in 1386, 'gathered with a wide embrace into his single heart, a sheaf of fatal Austrian spears', and thus opened a breach in the serried mailclad Austrian line through which the Swiss poured to score a great victory. Such great occasions of altruism are for the glorious history of patriotism. How can we poor mortals qualify? For myself, I hope that I qualify as a modestly altruistic person, yet I cannot imagine myself as being prepared to lay down my life even for my nearest and dearest.

So I will shift the discourse to a most mundane human activity – a family sitting down to a meal together. Such an example of food sharing is extremely rare in the animal kingdom. Food sharing is of course practised by the social insects such as bees and ants (Wilson, 1975), but it is a purely instinctive performance. By that I mean that genetic coding builds the central nervous system that programmes some stereotyped food sharing behaviour amongst the members of the insect colony. Except for the feeding of young birds and mammals the only examples of food sharing cited by ethologists (cf. Isaac, 1978) are the sharing of the prey caught by wild dogs, and to a very limited extent by chimpanzees. With these latter there is at most a grudging sharing of a large piece of meat. It is never offered, there being merely an unwilling tolerance of limited scrounging.

By contrast archeological evidence (Isaac, 1978), derived from detailed examination of Australopithicine sites of about 2 million years ago provides clear evidence that the food of the hunt was carried to the 'living site' for communal consumption. It was not eaten at the site of the catch. Thus these primitive hominids had already progressed from the selfish feeding habits not only of anthropoid apes, but of all their ancestors. Isaac suggests that this behaviour of some 2 million years ago was probably linked with division of labour and the beginning of culture. It was probably associated with the beginning of a language that was more than animal calls and cries. Certainly it was at the beginning of the rapid increase in brain size that seems most probably to be explained as due to the demands for increasing cerebral capacity to match the increasing linguistic performance. Of course this process of cerebral development is due to the natural selection very effectively given by the improved communication amongst members of the tribe (cf. Eccles, 1979a, Lecture 5). But this result of biological evolution, a propensity to speak, is linked with the creation of linguistic communication which is a World 3 phenomenon, belonging to cultural evolution. There is no way of obtaining information about the postulated linguistic development of hominids, but we can regard food sharing as indicative of an improved communication between members of a tribe. Food sharing of hominids could be the first example of true altruistic behaviour in the history of our planet.

It is surprising how sparse is the evidence of altruistic behaviour in hominids. However, with Neanderthals 80,000 years ago we have the first known examples of ceremonial burials, which certainly are altruistic acts. In human prehistory (60,000 years ago) the first evidence for compassionate behaviour has recently been discovered by Solecki (1971) in the skeletons of two Neanderthal men that were incapacitated by severe injuries. Yet the bones indicated that these incapacitated creatures had been kept alive for up to 2 years, which could have occurred only if they had been cared for by other individuals of the tribe. Compassionate feelings can also be inferred from the remarkable discovery that burials at that time in the Shanidar cave were associated with floral tributes, as disclosed by pollen analysis (Solecki, 1977). We thus may date the earliest known signs of altruism in human prehistory at 60,000 years ago. One could hope that it could be earlier because Neanderthal men with brains as large as ours existed at least 100,000 years ago.

With this historical background, how should we identify altruism in our present society? In the first place, it is essential to have *intent*; but the consequence is not definitive. *Intention* can be seen as one of the World 2 components in Fig. 1-7, and it was one of the themes in Lecture 4. Secondly, the intended action must be designed with *regard* for the interests of other persons, because altruism concerns only interpersonal relationships. It is evident that the ordinary social life of well meaning persons is a tissue of altruistic actions. Let us consider our own lives in families and in social groups. Let us even consider this lecture! I have tried hard to make it worth while for you to listen to, and you for your part are also behaving altruistically as a good audience and not throwing things at the lecturer! In every aspect of social life civilized behaviour is exhibited by the caring for others. This does not mean an uncritical demeanour. On the contrary criticism can and should be given in such a way that it could be of value to the recipient. In contrast to altruistic behaviour there is aggression that will be the theme of a later section.

In opposition to this simple human story of altruism is Chap. 7 of the book *On Human Nature* (Wilson, 1978), where Wilson distinguishes two basic forms of altruism which he calls hard-core and soft-core. Hard-core altruism seems to qualify as a pseudaltruism because he states that it is likely to have evolved through kin selection and serves selectively the altruist's close relatives. It is exhibited in its purest form with social insects. Wilson (1978) states:

> Human altruism appears to be substantially hard-core when directed at closest relatives, although still to a much lesser degree than in the case of the social insects and the colonial invertebrates. The remainder of our altruism is essentially soft. The predicted result is a melange of ambivalence, deceit and guilt that continuously troubles the individual mind.

So human altruism is largely of the soft-core variety, which is the altruism of traditional ethics. It is so distressing to read Wilson's account in this almost unbelievable diatribe:

> 'Soft-core altruism' ... is ultimately selfish. The 'altruist' expects reciprocation from society for himself or his closest relatives. His good behavior is calculating, often in a wholly conscious way, and his maneuvers are orchestrated by the excruciatingly intricate sanctions and demands of society. The capacity for soft-core altruism can be expected to have evolved primarily by selection of individuals and to be deeply influenced by the vagaries of cultural evolution. Its psychological vehicles are lying, pretense, and deceit, including self-deceit, because the actor is most convincing who believes that his performance is real.

Despite this devastating attack Wilson goes on to state:

> Human beings appear to be sufficiently selfish and calculating to be capable of indefinitely greater harmony and social homeostasis. This statement is not self-contradictory. True selfishness, if obedient to the other constraints of mammalian biology, is the key to a more nearly perfect social contract.

It would seem that Harvard is a sophisticated social hell and that the social insects are to be envied and admired! Wilson no doubt thinks that this is sociobiological realism. I think it is cynicism of a particularly obnoxious kind. Instead of being the most gracious of human values altruism is tainted with lying, pretense and deceit so that it is a monstrous hypocrisy; but in contrast Wilson states: 'There are no hypocrites among the social insects.'

When Wilson leaves the field of sociobiology and embarks on dogmatic statements on ethical issues, he opens himself to criticism on moral grounds. His statements are examples of the fundamentally flawed concepts of human nature that are developed by sociobiologists whose expertise is limited to non-human animals, and particularly to the social insects.

Under a heading 'The need for epistemic humility', Campbell (1975) states in reference to the debate on altruism and selfishness:

> The epistemic arrogance of behavioural and social scientists is perhaps as much an obstacle to understanding these matters as is the epistemic arrogance which traditional religionists exhibit in their claims of revelation and absolute certainty.

It is with great relief that I leave these murky platitudes of Wilson on altruism for the enjoyment of some poetic and imaginative ideas that Sherrington expressed in his Gifford Lectures at the University of Edinburgh in 1937–1938. In the last lecture of Man on his Nature (Sherrington, 1940, Chapter XII), he has been considering how altruism came into existence in the story of mankind. He emphasizes:

> The necessity of altruism for the future of humanity.... Man's altruism has to grow. It is not enough for him to stand and deplore. That is less to mend things than to run from them. A positive charity is wanted; negation is not enough. In effect it needs a self-growth which shall open out a finer self. It requires to absorb into 'feeling' something of the world beyond the self and put it alongside the interests of the very self. That asks biologically an unusual and even a perilous step.

In this account of the coming of altruism it is assumed that it was linked with the coming of self-consciousness, that is with the recognition of oneself and others as conscious selves. From this recognition would flow the caring for others when alive and the caring for the dead by the ceremonial burial customs that were practised by Neanderthals first about 80,000 years ago (Hawkes, 1965). The Neanderthal brain came at the end of an amazingly rapid development with a trebling of size over some 3 million years, which is an almost incredibly rapid evolutionary advance. Doubtless this cerebral capacity was required for linguistic performance, which would be of the greatest importance in welding a tribe together for hunting and warfare, as well as for social cohesion. However, the emergence of self-consciousness apparently came much later, after the brain had grown at about 100,000 years ago to a size matching ours. This emergence is referred to as a 'transcendence' by Dobzhansky (1967). Then came the first evidence of altruism in the ceremonial burial customs, though possibly it could be much earlier in the food sharing of hominids as described above.

Sherrington (1940) dramatically refers to the mysterious origin of morality and thus of altruism. Mother Nature, as exhibited in biological evolution apostrophizes man:

> You thought me moral, you now know me without moral. How can I be moral being, you say, blind necessity, being mechanism. Yet at length I brought you forth, who are moral. Yes, you are the only moral thing in all your world, and therefore the only immoral. You thought me intelligent, even wise. You now know me devoid of reason, most of me even of sense. How can I have reason or purpose being pure mechanism? Yet at length I made you, you with your reason. If you think a little you with your reason can know that; you, the only reasoning thing in all your world, are therefore the only mad one, ... You are my child. Do not expect me to love you. How can I love – I who am blind necessity? I cannot love, neither can I hate. But now that I have brought forth you and your kind, remember you are a new world unto yourselves, a world which contains in virtue of you, love and hate, and reason and madness, the moral and immoral, and good and evil. It is for you to love where love can be felt. That is, to love one another.
> Bethink you too that perhaps in knowing me you do but know the instrument of a Purpose, the tool of a Hand too large for your sight as now to compass. Try then to teach your sight to grow.

I have great empathy with my old master, Sherrington, in his deeply moving message. I believe that the biological evolution is not simply chance and necessity. That could never have produced us with our values. I can sense with him that evolution may be the instrument of a Purpose, lifting it beyond chance and necessity at least in the transcendence that brought forth human creatures gifted with self-consciousness. Cultural evolution then takes over from biological evolution and soon becomes crucial in natural selection, not only because of the wealth of technological innovations, but also because of the creation and development of the values. For example altruism serves well in giving the moral basis for a society dedicated to the welfare of its members.

In our modern world there are remarkable personal examples of altruism and compassion. I can instance Dr. Albert Schweitzer of Lamberene, Mother Theresa of Calcutta, and Dr. Guiseppe Maggi of the Cameroons. They have all dedicated their lives to caring for others who are in dire need of help. In a more communal way we have the many nations who are accepting the 'boat people' fleeing from the brutalities of Vietnam. Also many countries give asylum to individual people who are refugees from the slave states of the communist empires.

Now in my lecture I must turn to the realistically terrible threat by aggression. It has a biologically evolutionary origin and the propensity is built into the neural machinery of our brains, as has been considered in Lecture 5.

8.6 Aggression

At the outset it is important with Thorpe (1974) to distinguish between aggression and violence:

> We need two terms, *aggressiveness* in the normal biological sense and *violence* as behaviour definitely directed towards harming others. Above all we must avoid the cardinal error of assuming that in the higher animals and in man group aggressiveness is the necessary and inevitable result of hereditary constitution.

There is an enormous literature on human aggression. It has become an industry! We have been subjected to a spate of books, some with honest titles like Lorenz's *On Aggression* (1966), and Storr's *Human Aggression* (1968). Others have 'loaded' titles such as those of Morris, *The Naked Ape* (1967) and *The Human Zoo* (1969), Storr's *Human Destructiveness* (1972), Tiger and Fox's *The Imperial Animal* (1971), Ardrey's *The Territorial Imperative* (1966).

This doctrine of innate human aggression to an extreme degree has been poured out by this phalanx of writers who have either deliberately biassed the evidence or have been very limited and selective on the evidence they recognize. In Montagu's book (1976), *The Nature of Human Aggression*, there is a powerful attack on the extreme aspects of this doctrine of innate human aggression. One can hope that it will help to discredit the dogmas promulgated by those who believe in the intrinsic evil of mankind – almost beyond hope. Thorpe (1974) has expressed my thinking very well in his authoritative book *Animal Nature and Human Nature*.

> So far we have been using the term aggression quite loosely, but now we must be more precise, for an enormous amount of dangerously loose, sloppy thinking and prejudice is arising from just this source. It is highly regrettable that some of this has arisen from the recent writing of a very great zoologist, namely Dr. Konrad Lorenz. In his book, *On Aggression*, ... Lorenz commits the mistake of

> extrapolating too readily and uncritically from the behaviour of lower vertebrates such as fish and many birds to the behaviour of the higher animals and even man himself. Lorenz regards aggression as spontaneous and inevitably finding expression in violence, independently of external stimulation. ... But if aggression is not like that, we may be taking an unduly pessimistic view and resigning ourselves to the inevitability of the extreme form of aggression, namely war, when no such inevitability exists.

A myth promulgated by the innate aggressionists is the human pre-eminence amongst all mammals in exhibiting behaviour leading to violent conflict and death. Ethologists have now reported a greater aggressiveness in several species of mammals. For example Wilson (1975) states:

> On the contrary, murder is far more common and hence 'normal' in many vertebrate species than in man. I have been impressed by how often such behaviour becomes apparent only when the observation time devoted to a species passes the thousand-hour mark. But only one murder per thousand hours per observer is still a great deal of violence by human standards. In fact, if some imaginary Martian zoologist visiting Earth were to observe man as simply one more species over a very long period of time, he might conclude that we are among the more pacific mammals as measured by serious assaults or murders per individual per unit time, even when our episodic wars are averaged in.

Another myth of aggression is that our primitive ancestors, the Australopithicines, engaged in violent conflict so that most died with broken skulls (Dart, 1953; 1959; Ardrey, 1961). But as Montagu (1976, Chap. 6) has shown this is dubious evidence derived from fossil skulls that would have been damaged by natural causes. The same criticism is true also for the alleged violence of conflicts with the later hominids up to Neanderthal man. No doubt there was much violence, but the primitive hunter-gatherer peoples of today are little given to aggression and mostly live in harmony with their neighbours (Montagu, Chap. 7).

There are examples even today of peoples who are apparently non-aggressive. The Eskimos are a remarkable example. Another recently discovered example is the small isolated family of the Tasaday in Mindanao island of the Phillipines. There are 26 in all, 13 being children. I fear that they have been an anthropological exhibit since their discovery in 1971. No more can they have their simple isolated existence. I am reminded of the Australian joke about the primitive people in the highlands of New Guinea, where the standard family is father, mother, 3 children and 1 anthropologist!

Of course the historical record of mankind is replete with fighting and killing, but we are apt to overlook the unrecorded lives of the great bulk of the people that lived beneath all this surface of chivalric violence (cf. Eileen Power's book, *Medieval People*, 1951). The wonderful architecture from Romanesque times right through to the Rennaisance is a proof of the hard-working creativity of the people despite the veneer of violence. Chaucer's *Canterbury Tales* give a good example of humanity in the latter part of the fourteenth century. It is replete with good humour and coarseness, but not with violence.

In Lecture 5 I dealt with some of the structures at the base of the brain that are concerned with aggressive behaviour, the hypothalamus, the amygdala and the stria terminalis. As with the rest of the brain these are built by instructions from the genetic coding operating on the milieu of the embryo, and are fully formed soon after birth. However, full functional performance comes gradually during adolescence. The innate aggressionists would regard these structures as exercising a dominant role in human behaviour by providing drives for aggression and violence, as well as for sex, food and other pleasurable addictions. These structures are loosely spoken of by popular writers (cf. Sagan, 1977) as the reptilian brain on the erroneous analogy that reptilian behaviour is predominantly aggressive. It is too easy to think of the Tyrannosaurus, the alligator's jaws or the snake's venom. However the enormous majority of the 5000 species of reptiles are not aggressive, except in the search for food. Let us think of tortoises, turtles, lizards for example and let us abandon the myth of the reptilian brain as being at the root of human aggression.

I will now survey the case of human aggression from general principles. It is essentially the nature–nurture debate. Our brains are built with the inherited genetic instructions plus all manner of secondary influences in the developmental process. This statement can be made only as a general principle. The details are unknown. From birth onwards the *human being* is subjected to all the environmental influences of the culture, and so comes to be a *human person*. The significance of the environmental influence cannot be over-estimated, as I have described in Lecture 7 of my first Gifford series (Eccles, 1979a). Under the influence of a broken home environment and a sense of rejection a child can be alienated and be absorbed into groups that practise uncurbed aggression that spills over into violence.

There probably always has been some mesmeric attraction in spectacles of violence. I can read with revulsion of the Roman gladiatorial combats, which today continues in a less extreme form in such human aggressions as boxing. But now we have horror movies and horror programmes on TV that are a popular attraction. For example, Kubrick who produced the terribly violent, sadistic film *A clockwork orange* cynically states:

> I am interested in the brutal and violent nature of man because it's a true picture of him. Man isn't a noble savage; he's an ignoble savage. He is irrational, brutal, weak, silly ... any attempt to create social institutions on a false view of the nature of man is probably doomed to failure.

All that Kubrick is interested in are the profits from purveying brutality, evil and depravity.

As Hayek (1978) so well states:

> It is the harvest of those seeds that we are now gathering. Those non-domesticated savages who represent themselves as alienated from something they have never learnt ... are the necessary product of a permissive education which fails to pass on the burden of culture, and trusts to the *natural instincts which are the instincts of the savage.*

The TV and movies plus the pulp literature must be held responsible for the aggression and violence of our present society. I agree with Thorpe (1974) that the pornography of violence presents a greater danger than the 'pornography' of sex presented without violence or perversion. Yet because some beautiful human form is shown unclothed, a film is banned for minors by the censors, while the pornography of violence gets through to all ages.

It is a remarkable tribute to humanity that even with such terrible indoctrinations in violence our society is so peaceful. The academic mandarins who write these books on the depravity of mankind do not seem to realize the essential decency of the common people, as they would call them. I had an experience on August 14th of last year while having dinner outside a restaurant at the small seaside resort of Capo San Vito in Sicily. For almost 2 h there was a promenade, as if on a stage, of local people thronging the street. There were hundreds of families all dressed up in their best with children and perambulators for the occasion of the eve of the Assumption. As they promenaded backwards and forwards along the street closed to traffic, they would be greeting each other, conversing and laughing in the happiest way. The young people were joining into groups. The only unhappy person was the policeman who had nothing to do! Here is an example from our civilized society to match the stories of the simple hunter-gatherer peoples.

I present therefore the thesis that human beings are born with the potentiality to become fine human persons living in harmony with their fellow beings. Given a good home base and a good environment with their school fellows they can be relatively untouched by the horrors of much of the so-called entertainment industry, and by the subversive literature of Morris, Ardrey, Dart, Tiger and Fox and their followers. Of course, we all inherit brains with neural machinery appropriate for aggressive acts just as we have neural machinery for linguistic acts. However speaking is dependent on learning and this is also true of aggressive behaviour and violence. We can learn to control, which is the function of our great neocortex. As Wilson (1978) states:

> We must consciously undertake those difficult and rarely travelled pathways in psychological development that lead to mastery over and reduction of the profound human tendency to learn violence.

Of course I agree, but the 'rarely travelled pathways' are what is done by every mother in every good family.

I would add that in the learning of control 'sublimation' is most important. In this way we can turn our aggressive tendencies to good ends. Without a propensity for aggression I would not strive as I do for an achievement in accord with my hopes. For example, in composing these lectures I have been driven by my innate aggression to examine critically a wide range of literature, to synthesize it into meaningful stories, and to attack critically!

However, there is also a dark side to the future of human societies. As Sherrington (1940) in his last Gifford Lecture so well recognized, there is always the threat of *homo praedatorius* that can come in many disguises:

> Field after field of human civilized activity becomes a scene of conflict little less internecine than is war. Economic warfare, commercial warfare, class-warfare, are symptomatic of *homo praedatorius*. Serfdom and slavery attach to his régime, in fact when not in name. His régime deals by opprobrium and ostracism with victimized classes in the State. He exploits cruelty on sub-human lives as well as on human life. Predaceous man's rule cuts indeed at the very root of social mankind's organization of life.

Always there is the insidious threat from those who avidly seek positions of power and who are concerned with their own advancement rather than the altruistic motive of making the best contribution to the culture and well-being of their society. Today free societies are threatened as never before by terrorists ganged up from many countries. They may masquerade under some political guise, such as destroying the 'rotten' society in which they live in order to bring about some millenarian future. Many of the 'liberal left' have been blind to the dangers. But the criminal horrors of their murders, kidnapping and plane hijacking exposes them for what they are – criminal assassins exulting in destruction and killing for its own sake, and entirely barren of any constructive ideology. At last it is being recognized that there can be no negotiation with these criminals, who are a particularly vile form of *homo praedatorius*. They have to be fought to the death. Conor Cruise O'Brien in his article 'Liberty and Terror' (Encounter, October 1977), concludes: 'However long it takes, the democratic struggle against the underground terrorist empires, whatever slogans they may use, is a struggle for real liberty for real people'.

Meanwhile there has to be some curtailment of freedom in order to preserve our society with its values, which introduces us to the next lecture on the Quest for Meaning and Values.

Lecture 9

The Quest for Values and Meaning

Résumé

A simple definition of values leads on to consideration of their origin and development in cultural evolution. Special reference is made to the great Greek tradition. Though *absolute values* are beyond our attainment, it is argued that our highest endeavours are oriented to the search for absolute values in science and in the humanities, for example in the search for scientific truth and for ideal beauty.

In the modern world values are often given a purely material connotation – market values. The term 'innate values' with the brain operating as a value-driven decision system is comparable with pseudaltruism.

The theme of science and values involves a critical appraisal of Monod's claim that 'objective knowledge is the only authentic source of truth'. This claim is linked with a vicious attack on all traditional values and a denial of purpose anywhere in nature. It will be argued that Monod's dogmatism is based on a confusion of scientific thinking and on an emotional reaction against what he calls the *animist tradition.* As against Monod there are given the wise and deep statements of Heisenberg and Eddington on values, and reference is made to quotations by Popper and Planck.

It is shown that the common belief: 'science deals with facts, the humanities with values' is superficial and false. *Science is permeated with values, ethics in the search for truth and aesthetics in the conceptual judgement of hypotheses.*

The lecture closes with a consideration of two closely related themes: the purpose and the meaning of life. Recent books by Thorpe and Granit are devoted to purposive concepts in biology. Adaptability giving purposive behaviour is hailed as the great climax of evolution – *that purpose came into a world of chance.* Their sane outlooks contrast with the dogmatic scientistic statements in the last chapter of Monod's book.

Meaning is considered in relation to the present prevailing philosophy that questions of meaning should not be asked – or answered. Viktor Frankl gives the fullest expression to the urgency and importance of the quest for meaning in this age. The last lecture will be devoted to answering this challenge.

9.1 Origin of Values

Values can be thought of very simply to be what we value. They are at the basis of our judgements and our choices – whether to do this or that. Each of us has a scale of values or a value system that may not be consciously recognized, but which, nevertheless, provides a framework for our decisions; but of course, its role is to condition not to determine. Such rules of conduct are just as much learnt as is language. The brain is built with a propensity for language. Great linguistic areas of the neocortex have been developed in biological evolution and are at birth preformed before being used for whatever is the language that is heard. So also the brain is built with the propensity for performance according to value systems, and the value system initially learned is that of the ambient culture. Like language, value systems are important components of World 3.

Just as there are rudimentary animal communications, so there are rudimentary value systems, as can be seen by their social organization that may not be purely instinctive; or even by the rare altruistic performance of food sharing, as mentioned in Lecture 8. At the most we can speculate about the developing culture of early hominids (cf. Isaac, 1978) that was linked with the social cohesion, the food sharing, the developing language and the tool manufacture. It was in this way that values developed, for example in the products of tool manufacture with gradual improvement over hundreds of thousands of years. But we can conjecture that more importantly there was the gradual improvement in linguistic communication and in social organization.

As Sherrington (1940) reviews the origin of man he writes:

> We think back with repugnance to that ancient biological pre-human scene whence, so we have learned, we came; there *no* life was a sacred thing. There millions of years of pain went by without one moment of pity, not to speak of mercy. Its life innately gifted with 'zest-to-live' was yet so conditioned that it must kill or die. For man, largely emancipated from those conditions, the situation has changed. ... The change is in himself. Where have his 'values' come from? Those other creatures than himself, even the likest to himself, would seem without the values. There arises for him a dilemma and a contradiction. The contradiction is that he is slowly drawing from life the inference that altruism, charity, is a duty incumbent upon thinking life. That an aim of conscious conduct must be the unselfish life. But that is to disapprove the very means which brought him hither, and maintains him. Of all his new-found values perhaps altruism will be the most hard to grow. The 'self' has been so long devoted to itself as end.... Man is grappling with its newly found 'values', yet with no experience except its own, no counsel but its own.

As Hayek (1978) so succinctly states:

> Man has certainly more often learnt to do the right thing without comprehending why it was the right thing, and he still is often served better by custom than understanding. ... Thus a tradition of rules of conduct existing apart from any one individual who had learnt them began to govern human life.

In Lecture 6 of the last Gifford series (Eccles, 1979a) I outlined the cultural story of mankind through to the first great civilization – the Sumerian. Already there had been the first document of human rights, the Code of Ur-Nammu at about 2100 B.C., and also superb artistic, literary and technological developments. Presiding over the complex creative society of a Sumerian city state with their code of values was the priestly bureaucracy of the Ziggurat deriving its authority from the god or system of gods whose deliverances were interpreted by the priests into rules of conduct of the affairs of the people of the state. It has recently been proposed by Julian Jaynes (1976) that the priests of the Ziggurat were not charlatans, but that, when in hallucinatory trances, they heard voices which they genuinely believed to be the voices of the god!

In retrospect we can recognize that the societies of the early civilizations were governed by systems of values that were prescriptions of conduct and of judgement. Moreover these systems were slowly evolving and just as today were very different for the different cultures – Egyptian, Sumerian, Assyrian, Indian, Chinese. As Hayek (1978) states with respect to cultural evolution:

> If we are to understand it, we must direct our attention to that process of sifting of practices which sociobiology systematically neglects. It is ... the most important source of what ... I have called human values.

It appears that this pervasive cultural milieu was unconsciously accepted without critical examination. The conscious recognition of values seems to have begun with Socrates who persisted in asking new questions relating to the value of knowing, to the purposes of such activities as science, politics, the practical arts and even of the course of nature. These questions led to the concept that everything had a value, that is to the general notion of value. As Susanne Langer (1951) states so eloquently:

> That everything had a value was too obvious to require statement. It was so obvious that the Ionians had not even given it one thought, and Socrates did not bother to state it; but his questions centered on what values things had – whether they were good or evil, in themselves or in their relations to other things, for all men, or for few, or for the gods alone. In the light of that newly-enlisted old concept, *value*, a whole world of new questions opened up. ... What is the highest good of man? Of the universe? What are the proper principles of art, education, government, medicine?... Why have we passions, and a dream of Truth? Why do we live? Why do we die?

With this background Plato developed his ideas of essences or ideals of which the most important are: the Good; the Beautiful; and the True. They were conceived to be eternal, of divine origin and to be the Absolutes against which all values were judged. It must be recognized that this Platonic Third World is completely different from the World 3 of Popper (Lecture 7) in that it is of divine origin, whereas the Popperean World 3 is entirely human in origin. It is also different in respect of its absolute status. According to Plato if anything else apart from the idea of absolute beauty is beautiful, then it is beautiful *for the sole reason* that it has some share in the idea of absolute beauty. *And this kind of explanation applies to everything* (cf. Popper, 1972). This idealist trinity of absolute values survived in the idealist philosophies of recent times. A lovely variant is in the last lines of Keats' 'Ode on a Grecian Urn':

> 'Beauty is truth, truth beauty', – that is all Ye know on earth, and all ye need to know.

Although the World 3 of Popper is entirely of human origin it can approach to an absolute value. For example, Popper (1972) states with respect to the methodology of science:

> Thus the elimination of error leads to the objective growth of our knowledge – of knowledge in the objective sense. It leads to the growth of objective verisimilitude; it makes possible the approximation to (absolute) truth.

Popper regards science as a *search* for absolute truth, which though unattainable is the goal and criterion of our efforts. In contrast Hegelian relativism is to be rejected because it leads to spurious scientific truths based upon ideologies like Marxism. Popper (1975) emphasizes the role of tradition:

> The rationality of science consists in the demand that a new theory must transcend its predecessors by preserving their successes. My thesis, therefore, is that science is based, essentially, on tradition; that it is a two-level tradition – conservative as well as revolutionary. So the critical tradition has a two-storey structure; the first-order tradition consists of a record of what our predecessors have said, of their various theories; the second-order tradition consists of adopting a critical view of the first-order tradition. ... The ethics of science has to be founded in a tradition, and I think that the loss of this tradition would mean the real end of our civilization. Among the various elements constituting the ethics of science are the respect for truth and the respect for the impersonal character of critical discussion.

The great physicist Max Planck (1950) made a striking affirmation of his belief in absolute values:

> I emphasized that I had always looked upon the search for the absolute as the noblest and most worth while task of science. ... These absolute values in science and ethics are the ones whose pursuit constitutes the true task of every intellectually alert and active human being. ... This task is never finished – a fact guaranteed by the circumstance that genuine problems ... constantly appear in ceaseless variety and constantly set new tasks for active human beings. For it is work which is the favourable wind that makes the ship of human life sail the high seas, and as for the evaluation of the worth of this work, there is an infallible, time-honoured measure, a phrase which pronounces the final authoritative judgement for all times; *By their fruits ye shall know them!*

9.2 Values in the Modern World

A greatly changed attitude to values can be recognized in a comprehensive symposium entitled *Values and the Future* (Baier and Renscher, 1969). This selective examination of values was concentrated on economics, technology and political issues, because these were more susceptible to the 'measurability required if a science is to be built'. I make a brief reference because it illustrates the limitations often put on values in the modern world. For example Baier states:

> That I believe or value something *consists in* my having a certain attitude or behavioural disposition. I am disposed to act on my belief, to take protective action in defence of what I value. ... But *what* I believe or value is dependent on the relevant appraisals. ... I do not value things for whose valuelessness I have overwhelming evidence.

Linked to that concept is his earlier statement:

> The concept of value, in the sense used in 'the value of a thing' is central to traditional economic Value Theory. The value here investigated is of course the so-called exchange or market value of a commodity.

If we value possessions we have a dedication to 'market values' which lies at the basis of the free economy.

A more generalized treatment of values is given by Pugh (1978) in a comprehensive and ambitious monograph entitled: *The Biological Origins of Human Values.* Corresponding to the innate altruism of sociobiologists which I called pseudaltruism in Lecture 8 it is proposed by Pugh that there are innate values which are the products of evolutionary design and that 'are built into the hardware of the modern human brain ... a production of evolution and genetics'.

The list of innate values includes the unpleasant – pain, hunger, thirst, sorrow, shame, fear, anger – and the pleasant – comfort, tactile pleasure, good taste, good smell, joy, pride and sexual experience. It is certainly a mixed bag and would do quite well for a modern chimpanzee as well as for a modern human being! As we would expect, Pugh is a devotee of Sociobiology. My comment is that the term 'innate value' is a misuse of the concept of value. *There are no genes for the values that matter for human living. These values were developed in our cultural evolution, biological evolution contributing the brain with its potentialities.* The realization of these potentialities is due to the educational process.

In the greater part of his book Pugh speaks of values according to the decision-theory perspective. The human brain is a value-driven decision system that was evolved in biological evolution. The claim is made that this book 'for the first time places the study of values on a firm scientific foundation'. These statements are of course for the innate values which I regard as on a par with pseudaltruism (cf. Lecture 8). Fortunately Pugh is very uncertain about how far intellectual values are in the innate category. For example he is doubtful if there is any innate value associated with truth, though he thinks that 'instinctive values' influence our appreciation of the arts.

In the last part of his monograph Pugh refers to values as they are ordinarily understood as: 'secondary values which are products of rational thought and therefore are subject to change on the basis of rational thought'. In this section he has left his arid design systems theory with innate values and writes about values as a dedicated humanist:

> Some of the most important human values are emotional rather than intellectual. If we become too narrowly intellectual and analytical in our personal lives there is a risk of shackling emotions, and numbing the zest for life. To gain the full benefit of the theory as a social concept and a personal philosophy it will be necessary to explore the symbolic and emotional side of the theory. This final shift of perspective cannot be accomplished by a scientific approach. It requires emotional and artistic techniques; and it is a task for creative artists, writers, musicians, and moral leaders. ... We must not allow the scientific answers to spoil the philosophical questions. From a subjective and emotional perspective the fundamental mysteries of human existence remain as a profound personal enigma – with the result that we enjoy the ambiguity and mystery of literary and poetic speculation. ... The modern educational emphasis on the scientific method, combined with the deemphasis of spiritual, religious, and philosophical concepts, has produced a distortion in our culture in which 'rational' materialistic values are exaggerated and the spiritual and emotional values are largely ignored or suppressed by the 'rational' mind.

Pugh is a deeply concerned and courageous thinker. I believe that he has overstressed the design systems approach in the first section of the book, which is his effort to develop a radically new approach to values. His criticism of educational practices is most timely. The teaching of moral values to the young is often derided as indoctrination or even as brain-washing. Would these critics also make the same criticisms of the teaching of a language? I would maintain that it is a crime against children to leave them to their own devices in learning a language. They can therefore suffer from a life-long deficiency. The wonderful genetic building of a brain for all human expressions in language is left derelict, for that is what happens if it is not intensively developed in the formative years. The same is true for moral education. The impressionable brain of the child has to be given moral instruction otherwise that child is likely to grow into an adult permanently maimed or flawed in human qualities. It is as much a crime as to bring children up with an inadequate linguistic performance.

9.3 Science and Values

We can regret that Monod's great scientific achievements and the excellent scientific chapters of his book *Chance and Necessity* were confused by and mixed with an extravagant ideology. As scientists we must take note of the vicious attack made by Monod (1971) on all systems of values. Thus he states:

> The cornerstone of the scientific method is the postulate that nature is objective. In other words, the *systematic* denial that 'true' knowledge can be got at by interpreting phenomena in terms of final causes – that is to say, of 'purpose',

which he refers to contemptuously as an 'animism'. This dogmatism will be criticized later. The argument goes on with vehemence in the apocalyptic last chapter that is appropriately named 'The Kingdom and the Darkness'. I must quote at some length these almost unbelievable diatribes against all types of modern society, including both liberal and Marxist:

> In both primitive and classical cultures the animist tradition saw knowledge and values stemming from the same source. For the first time in history a civilization is trying to shape itself while clinging desperately to the animist tradition to justify its values, and at the same time abandoning it as the source of knowledge, of *truth*. For their moral bases the 'liberal' societies of the West still teach – or pay lip-service to – a disgusting farrago of Judeo-Christian religiosity, scientistic progressism, belief in the 'natural' rights of man, and utilitarian pragmatism. ... However this may be, all these systems rooted in animism exist at odds with objective knowledge, face away from truth, and are strangers and fundamentally *hostile* to science, which they are pleased to make use of, but for which they

> do not otherwise care. The divorce is so great, the lie so flagrant, that it afflicts and rends the conscience of anyone provided with some element of culture, a little intelligence and spurred by that moral questioning which is the source of all creativity. ... For behind the protest is the denial of the essential message of science. The fear is the fear of sacrilige: of outrage to values. A wholly justified fear. It is perfectly true that science outrages values. Not directly, since science is no judge of them and *must* ignore them; but it subverts everyone of the mythical and philosophical ontogenies upon which the animist tradition, from the Australian aborigines to the dialectical materialists, has made all ethics rest: values, duties, rights, prohibitions.

We may note in passing that Monod having 'blasted' values and morality dares to appeal to 'conscience' and 'moral questioning'! Having thus abandoned values of all types of societies, all being tainted by animism, Monod attempts in the last pages of 'The Kingdom and the Darkness' to salvage his dogma that 'objective knowledge is the only authentic source of truth', by the contradictory arguments on page 176. He argues that:

> true knowledge ... cannot be grounded elsewhere than upon a value judgement. ... It is obvious that the positing of the principle of objectivity as the condition of true knowledge *constitutes an ethical choice and not a judgement arrived at from knowledge, since, according to the postulate's own terms, there cannot have been any 'true' knowledge prior to this arbitral choice. ... Hence it is from the ethical choice of a primary value that knowledge starts.*

Thus in these bewildering tactics the value judgement of ethics has to be introduced momentarily for the arbitral choice of true knowledge, yet we are told just before that: 'ethics ... is forever barred from the sphere of knowledge'.

In his review of *Chance and Necessity* in *The Times*, Thorpe, (May, 11, 1972) points out this logical *inconsistency*, stating that he felt 'menaced by a vague spectre compounded of Teilhard de Chardin, D.H. Lawrence and J.P. Sartre!'

Thorpe (1978) makes a further comment:

> Monod and I avowedly set out guided by the same fundamental belief that to attempt to understand the natural world in the way in which science attempts the task is inherently worth doing ... we believe that the results, if achieved by rigorous and cogent thought, will have their own intrinsic value for mankind. This 'value' implies what I call 'natural religion' or a 'metaphysic' though Monod would have used neither of these terms.

I will now criticize the dogmatic statement of Monod (1971) that:

> Science as we understand it today ... required the unbending stricture implicit in the postulate of objectivity – ironclad, pure, forever undemonstrable ... of the *nonexistence* anywhere in nature of a purpose, of a pursued end.

We may ask how can this dogmatism be presented in the face of the many counter-examples that are provided by great scientists of the twentieth century who seemed to do quite well in science despite their belief in some meaning or purpose in existence, and particularly in biology and in human existence, and who believed in values and ethics. I give herewith a list of references to their publications on these themes, but more detailed reference to several have been given in Lecture 1 of my first Gifford series (Eccles, 1979a) and will also be given later in this lecture and in the next: Planck (1937, 1950); Schrödinger (1958); Heisenberg (1971); Eddington (1939); Hinshelwood (1961); Polanyi and Prosch (1975); Sherrington (1940); Thorpe (1962, 1974, 1978); Wigner (1964, 1969); Sperry (1975, 1976).

Monod does not seem to know the published testimonies of these great scientists. They practised science wihout recognizing 'the non-existence anywhere in nature of a purpose, of a pursued end' that Monod makes the cornerstone of the scientific method. This is evidence sufficient for rejecting his dogmatic claim. Some of these references postdate *Chance and Necessity*, but all authors had earlier publications.

Cournand and Meyer (1976) severely criticized Monod on the grounds that he neglected so many of the fundamental obligations of the scientist in his obsession with the obligation of objectivity. Furthermore they state that Monod's

> argument if accepted does not imply the need to give up 'animist' values, but only to recognize that they are not derived from scientific knowledge. He has not demonstrated their incompatibility with and 'hostility' towards science.

For this I am grateful because Monod kindly accused me on several occasions of being an animist!

Skolimowski (1974) criticized the incoherence of Monod's position with respect to his restriction of science by the principle of objectivity and points out how great scientists of the past believed in a Master Plan – Copernicus and Newton for example. Some confusion has arisen over Monod's specialized usage of the word 'objectivity' (cf. Monod, 1974). In an attempt to prevent further confusion I will reinstate objectivity with its customary philosophical meaning by giving a succint statement by Popper (1972):

> Only objective knowledge is criticizable; subjective knowledge becomes criticizable only when it becomes objective. And it becomes objective when we *say* what we think; and even more so when we *write* it down, or *print* it.

Let us now return to the same tradition of values that we experienced in the earlier quotations from Popper (1972, 1975) and Planck (1950). Heisenberg (1971) has presented wise and deep statements on values:

> The problem of values is nothing but the problem of our acts, goals and morals. It concerns the compass by which we must steer our ship if we are to set a true course through life. The compass itself has been given different names by various religions and philosophies; happiness, the will of God, the meaning of life – to mention just a few. The differences in the names reflect profound differences in the awareness of different human groups. I have no wish to belittle these differences, but I have the clear impression that all such formulations try to express man's relatedness to a central order. ... In the final analysis, the central order, or the 'one' as it used to be called and with which we commune in the language of religion, must win out. And when people search for values, they are probably searching for the kind of actions that are in harmony with the central order, and as such are free of the confusions springing from divided, partial orders. The power of the 'one' may be gathered from the very fact that we think of the orderly as the good, and of the confused and chaotic as the bad.

The great philosopher-physicist Eddington (1939) has expressed beliefs comparable with those of Heisenberg. The possibility of the arbitrariness of values derived in human consciousness led him to a belief in absolute values and that we can, 'trust optimistically that our values are some pale reflection of those of the Absolute Valuer'.

While on the subject of science and values, it is important to expose and reject a widely held belief that science is value-free. 'Science deals with facts, the humanities with values'. This is a fundamentally mistaken view of science, which is the great enterprise of trying to understand all natural phenomena – inorganic, biological and psychological. Scientific theories in themselves are not facts.

They belong to World 3. Strictly speaking the role of facts in science is in the testing of theories and as background in the development of new theories (Popper, 1959). As mentioned in Lecture 1, Sperry has for many years advocated the view that there is a mutual interaction of conscious phenomena and brain processes, the mental phenomena transcending the physiological. Thus Sperry (1975) states:

> Subjective values of all kinds, even aesthetic, spiritual and irrational come to be recognized as having causal control potency in the brain's decision-making process – along with all other components of the world of inner experience ... all mental phenomena including the generation of values, can be treated as causal agents in human decision making. ... If judgements about right and wrong are best arrived at on the basis of what is true, avoiding what is false, science would seem on this count as well to deserve a leading role in determining ethical values instead of being disqualified.

Finally he concludes that: Science deals with values as well as with facts.

9.4 Conclusions with Respect to Values

It is not possible to summarize the diverse views or values that have been quoted above, but instead I will try to distil from it my philosophical position, which is based on the great tradition of values as participating to some extent in the absolute ideal, and which we have seen in the quotations to be continued into the twentieth century by Sherrington, Popper, Planck, Heisenberg and Eddington.

The cultural achievements of mankind bear witness to the search for absolute values that has motivated and inspired the great creative geniuses. It can be said that, symbolically, absolute values have provided a guiding beacon light. This can be appreciated when we consider the scientific efforts of Kepler, Newton and Einstein to understand the natural world. A similar guidance of geniuses can be discerned in other fields of cultural achievement: philosophy, religion, literature, history and the arts. The thoughts and aspirations of mankind in respect of truth, goodness and beauty have led to the search for justice and for codes of ethics in social organization.

The traditional values of truth, goodness and beauty are intertwined in the great cultural achievements and are used as criteria of judgement. No claim should be made that absolute values have been attained in any human enterprise. For example, the aim of scientists is to reach a deeper and wider understanding of natural phenomena by hypotheses of utmost generality, but no claim must be made that some hypothesis is an absolute truth. It is no more than the nearest approach to truth that we have yet achieved, and doubtless it will be superseded. In this sense the scientific enterprise can be regarded as a search for absolute values, not only of truth but also of beauty. Do we not recognize the beauty as well as the truth of some wonderful new insight of great generality and simplicity, as for example the double helix of the genetic code?

It is important to recognize that science is not restricted to phenomena that can be measured. On the contrary the identifying characteristic of science is its reliance on creative imagination and rational criticism. The harmony among the sciences derives from their common metaphysical principles: namely that creative imagination is exercised in attempting to develop hypotheses that are in conformity with existing knowledge and that await challenge by new evidence discovered by research. Always the aim is to approach closer and closer to absolute truth.

Science is permeated with values – ethics in our critical efforts to arrive at truth and aesthetics in our conceptual imagination and in the appreciation of our hypotheses. If we can give to mankind an understanding of science as a very human endeavour to understand nature and ourselves, then science would be appreciated as a great and noble human achievement, whereas, instead, it is in danger of becoming some great monster feared and even worshipped, but carrying the threat of destruction of humanity.

9.5 The Meaning and the Purpose of Life

These two words are closely related. 'Meaning' has a special reference to the significance and to the values of life, whereas 'purpose' is more specific, particularly in relation to the functioning of organisms. When materialist-behaviourist philosophy dominated biology, any reference to purpose was rejected as savouring of teleology with its forbidden implications that life was purposive in relation to final causes. Even Monod (1971) had to revolt against this restriction with his use of the word 'teleonomy' instead of teleology. And now we have two recent books by distinguished biologists with the word 'purpose' in the title – *The Purposive Brain* by Granit (1977), and *Purpose in a World of Chance* by Thorpe (1978).

Granit (1977) makes wise and witty comments:

> I am quite satisfied with purposiveness because, as we shall see, many nervous acts definitely have a purpose. ... Clearly it is often possible and even desirable to neglect purposiveness in studying established neural events from the physico-chemical point of view. It may well happen that the purpose for a time disappears into midair like the grin of a Cheshire cat. However, if one then tries to stick the pieces together but does it in a teleologically illogical manner, the cat again materializes, now grinning derisively at the bungler who is missing the point. We are in fact dealing with two ways – the why and the how – of approaching the same neural event.

It is relevant now to refer to part of a quotation from Sherrington (1940) that I gave in Lecture 8. He spoke of Mother Nature as, 'the instrument of a Purpose, the tool of a Hand too large for your sight as now to compass. Try then to teach your sight to grow'.

Monod (1971) attempts to bring meaning into his existentialist world of science with its authentic discourse based on the 'ethic of knowledge':

> Man ... progressively freed both from material constraints and from the deceitful servitudes of animism ... could at last live authentically, protected by institutions which ... could be designed to serve him in his unique and precious essence. A utopia. Perhaps. ... The ancient convenant is in pieces: man knows at last that he is alone in the universe's unfeeling immensity, out of which he emerged only by chance. His destiny is nowhere spelled out, nor is his duty. His kingdom above or the darkness below: it is for him to choose.

The messianic motive of the 'kingdom' worries me. It sounds like an elitist realm where ordinary people have no place or meaning. But my strong aversion is aroused by this pervasive existentialist dogmatism. I would demand the right to choose neither Monod's kingdom nor the 'other place'! It seems that 'the deceitful servitudes of animism' are much more attractive with their opportunities for living in fulfilment with love, friendship, art, music, literature; and with the hope that it is not 'by chance', but that we are participants in some great design with transcendental meaning, which was the theme of my last Gifford series – 'The Human Mystery'. Monod has no scientific right to say 'by chance'. It is a cover up of his ignorance.

Thorpe (1978) criticizes Monod's utopia on similar lines:

> What he does ... is to propose that objective scientific knowledge should replace religion, not only as a source of knowledge of the world, but also as a source of authority which determines the whole of man's being, even his innermost feelings and aspirations.

The utopia is exposed as a nightmare of scientism.

Polanyi and Prosch (1975) in a book published after Polanyi's death make important statements of meaning:

> If we believe in the existence of a general movement towards the attainment of meaning in the universe, then we will not regard any of the kinds of meaning achieved by men to be merely subjective or private. ... We will, on the contrary, regard every achievement of any sort of meaning as the epitome of reality, for we will think it is the sort of thing that the world is organized to bring about ... we will wish to live in a society in which their attainment is honoured and respected. ... We will not, for instance, want to see them controlled in some supposed public interest.

Dubos (1968) in his book *'So Human an Animal'*, expresses very well the alienation of today:

> The most poignant problem of modern life is probably man's feeling that life has lost significance. The ancient religions and social creeds are being eroded by scientific knowledge. ... As a result, the expression 'God is dead' is widely used in both theological and secular circles. Since the concept of God symbolized the totality of creation, man now remains without anchor. ... The search for significance, the formulation of new meanings for the words God and Man, may be the most worthwhile pursuit in the age of anxiety and alienation.

I am sympathetic to Dubos' expression, but I believe that there has been an unfortunate panic in many theological circles. Scientific knowledge leads to a natural theology that gives great religious opportunities. The universe is being recognized as more wonderful than has ever been realized, with unique design not only in the initiating 'Big Bang', but also in the chain of contingency that led to our existence here and now, as I described in my last Gifford series (Eccles, 1979a). I will develop this theme in the final lecture.

The urgency and importance of 'Meaning' has been expressed with great cogency and feeling by Dr. Viktor Frankl (1969). He is a distinguished Viennese psychiatrist who survived a concentration camp in which his family perished.

> It is an inherent tendency in man to reach out for meanings to fulfil, and for values to actualize. But, alas, we are offered by two outstanding American scholars in the field of value-psychology the following definitions: 'Values and meanings are nothing but defence mechanisms and reaction formations'. Well, as for myself I am not willing to live for the sake of my reaction formations, even less to die for the sake of my own defence mechanisms. And I would say that reductionism today is a mask for nihilism. ... This situation accentuates a world-wide phenomenon that I consider a major challenge to psychiatry – I have called it the existential vacuum. More and more patients are crowding our clinics and consulting rooms complaining of an inner emptiness, a sense of total and ultimate meaninglessness of their lives ... We may define the existential vacuum as the frustration of what we may consider to be the most basic motivational force in man, *the will to meaning* – in contrast to the Adlerians' will to power and to the Freudians' will to pleasure ... and actually man is not concerned, not primarily concerned with pleasure or with the so-called pursuit of happiness. Actually, due to his will to meaning, man is rèaching out for a meaning to fulfil or another human being to encounter; and once he has fulfilled a meaning or has encountered another human being, this constitutes a reason to be happy. ... Whereas, if this normal reaching out for meaning and beings is discarded and replaced by the will to pleasure or the 'pursuit of happiness', happiness falters and collapses; in other words, happiness must ensue as a side-effect of meaning fulfilment.

This conclusion can be related to Lecture 5. The full gratification of the emotional brain does not satisfy the human spirit. These wise and inspired thoughts of Frankl lead me to conclude this lecture by some general statements on human values and meaning. For me one of the essential ethical postulates is that we should be free. Each human being is a unique self with potentialities that give wonderful promise with all their great diversity. The ideal is for each human being to have the maximum freedom to realise its potentialities. This ethic derives from my belief that life has a transcendental meaning and that each life is precious. Together we are engaged in the tremendous adventure of consciously co-experienced existences. So we come to the themes of the next lecture that relate especially to the religious view of life.

Lecture 10

The Psyche, Freedom, Death and Immortality

Résumé

At the outset there is an attempt to bring together the main themes of the lectures. The hypothesis of dualist-interactionism is shown to have great explanatory power for this wide range of phenomena.

The use of the word *psyche* for soul in the theological sense is introduced by reference to Socratic statements in the *Phaedo* and the derivative comments by Hick. This dualism of the Socratic-Platonic philosophy leads on to the strong dualism of Descartes at the outset of modern philosophy. In the light of modern neuroscience body–mind dualism must be transformed into brain–mind dualism or *brain–psyche dualism*.

The important theme of the uniqueness of the psyche leads on to a religious view of the human person. There is firstly the uniqueness of each psyche that is not explicable on any scientific grounds, even by the most subtle genetics and neuroembryology. The coming-to-be of each experiencing psyche is mysterious, as was recognized long ago by Jennings. Thus we are led to a *creationist doctrine* of the origin of each unique human psyche. This theme is critically explored in all of its subtlety in relation to the hypothesis of dualist-interactionism.

During life our most common experiences are that we can at will bring about so-called voluntary actions, or that we can direct attention or recall memories. It is claimed that all materialist theories of the mind ultimately are reducible to determinism (cf. Lecture 1), and hence lead to a denial of human freedom and rationality. In contrast dualist-interactionism is consonant with a life engaged in the unending quest for the highest values – truth, goodness and beauty – that gives purpose and meaning to life; and for the freedom that entails moral responsibility. But necessarily this philosophy raises great unsolved problems.

It is important to recognize the fallacy of the prevailing belief that science has destroyed religion. Quotations from Heisenberg, Hinshelwood, Torrance, Einstein and Planck refute the superstition that science has completely replaced religion and that 'God is dead'.

On the dualist-interactionist hypothesis the psyche exists as an independent entity at the core of World 2 in liaison with the brain in World 1. The ultimate question is: what happens at death of the brain? Is the psyche then also completely annihilated or can we have hope that there is some quite unimaginable future? In this context there is discussion of the key problem: can there be *self-recognition after death?* All detailed memory must be lost, but it is suggested that a more general memory is associated with the psyche or World 2, and this could carry self-identity.

In the Epilogue there is reversion to the Great Questions relating to *purpose and meaning of our life* with all of the wonders and beauties of nature and of human experiences and creativities. In the context of the Natural Theology developed in these lectures, it is suggested that there is some *supernatural meaning* in the great design and in the great drama that is revealed in cosmology, in life and evolution and in the eventual coming-to-be of self-conscious beings in World 2 that have created another world, World 3.

The lectures of these two Gifford series thus can be seen to be part of a wide-ranging enquiry into existence in order to discover if there is purpose and meaning at a transcendental level.

10.1 Introduction

At the outset of my final Gifford Lecture, I will survey some of the main themes of my course.

In Lecture 1 there was a rejection of all the materialist theories of the mind and a brief statement of the alternative theory, dualist-interactionism. This theory was fully developed in Lecture 2 in relation to the recent discovery of the organization of the association neocortex into anatomical units or modules. It was an extension of the strong dualist-interactionism first proposed in the book by Popper and Eccles (1977). In a detailed evaluation of this theory it was shown that the organization of the neocortex into about 4 million modules could give it a capacity to generate an almost infinite number of spatio-temporal patterns – even enough for providing the neural substrate for our whole lifetime of memories and experiences. The liaison between the mind (World 2) and the brain in World 1 would be effected by the reciprocal interaction (cf. Fig. 1-7) between mental events and the neural events in the modules. An essential postulate of this strong version of dualist-interactionism is the existence of some conscious experiences *prior* to the appearance of the counterparts in the specific modular patterning in the neocortex. However the action time might be extremely short. If it were zero, then dualist-interactionism could be reducible to a variety of parallelism such as the identity theory. The mental events would have no *causal* relationship to the brain events.

By contrast on the perceptual side to the left of Fig. 1-7 and also with some items on the right – feelings and dreams for example – the brain events would precede the perceptual experiences, whereas on all materialist theories of the mind the relationship would be simultaneity. The perceptual interval is certainly less than 0.1 s. The longer values – up to 0.5 s in Libet's (1973) experiment – are largely occupied in the building up of modular patterns to the threshold of conscious experience. The best evidence for the duration of the perceptual interval is that derived from the recognition of depth perception which is about 40 ms in the Julesz experiments (Lecture 3). With the switching time in ambiguous figure recognition (Lecture 3), the perceptual time also must be very brief.

On the right side of Fig. 1-7 some items involve initiation by the World 2 event of subsequent neural events, e.g. in intention and attention. As described in Lecture 4, a conscious decision acts upon the brain to bring about modular activities that generate electrical potentials, the N 140, P 400 waves with attention, and the readiness potential triggered by an intention. Also the increased cerebral activity results in an increased circulation rate through the active areas (Figs. 6-1, 6-3). With the readiness potential the action-time of mind to brain has not been measured, but would certainly be less than 0.1 s. It is important to recognize that the 0.8 s duration of the readiness potential (cf. Fig. 4-9) relates to the cerebral events *following* the initiating intention. However it can be assumed that the intention is not a brief trigger event, but that it continues for the whole duration of the readiness potential and aids in the growth of the neural events and in their guidance onto the appropriate pyramidal cells of the motor cortex. It is also the continuing intention that gives the neural events during the goal-directed movement (Fig. 4-10). Another important mind to brain operation occurs in the initiating of a memory. The signal to search for a memory in the data banks of the brain comes from World 2, a conscious decision, to the liaison brain (cf. Fig. 1-7). The memory delivered to the conscious mind is there evaluated for correctness by what I have called the recognition memory, as described in Lecture 7.

It has been shown in these lectures that dualist-interactionism has great explanatory power in respect of many phenomena that have otherwise not been explained. For example the unity of conscious experience (Lecture 1) cannot be derived from brain activities in themselves, which remain disparate unless brought together in the mental synthesis. This is particularly evident in visual perception, where the unity of the observed picture is retained despite the saccadic movements of the eyes (cf. Fig. 3-5), which would result in a wild tumult of neural events in the visual cortex. There has been no attempt at explaining the stability of the visual experience in terms of materialist theories of the mind, whereas, in contrast, on dualist-interactionism the steady visual experience is attributed to the learned skill of the mind in scanning and synthesizing the disparate cerebral events (cf. Fig. 2-10).

In Lecture 5 dualist-interactionism provided in principle a basis for explanations of emotions and pain in all their various manifestations. In Lecture 6 dualist-interactionism was the basis of plausible explanations for the levels of consciousness obtaining for a wide range of cerebral activities from seizures to coma, and also for surgical anaesthesia. However the unconsciousness of sleep provides an enigma – as it does for all other brain–mind theories. Here is a challenge for future research.

It is in Lecture 7 that the independence of World 2 becomes so crucial in our efforts to understand World 3, remembrance and creativity. A strong dualist-interactionism meshes in well with these higher levels of human activity. The great explanatory power of dualist-interactionism is revealed.

Finally in Lectures 8 and 9 altruism and aggression, values and meaning are all illuminated and so become more clearly understood by being exposed to the searchlight of dualist-interactionism. It was there shown that purely materialistic theories of the mind result in confusions and misunderstandings, and in particular fail to discriminate adequately between the distinctively human and the animal performances.

This last lecture moves ultimately into the intimidating field of death and immortality. It is important to recognize that on any materialist theory of the mind this lecture would come to an abrupt end with death. The brain disintegrates along with any of the mental attributes it had on any materialist theories, the panpsychist, the epiphenomenalist or the parallelist in all of its various categories as enumerated in Lecture 1. Lewis (1978) refers to all these materialist philosophies as the corporealist view of the person. With all of them nothing of the person survives the death of the body. That is not to say that the alternative theory of dualist-interactionism ensures us of a spiritual immortality. At the most it gives hope, but not assurance (Lewis, 1978; Hick, 1976; Badham, 1976). The exploration of this problematic situation will come later in this lecture.

10.2 The Psyche (or Soul)

In this lecture series I have not so far introduced the Greek word 'psyche', which means soul in religious discourse. I have in general not used soul because this word has become so debased as to be almost derisory in such usages as: save our souls, soul music, soul food, etc. Hick (1976) in fact approves of this debasing of 'soul' because he believes that in religious discourse we should talk of 'persons' and not of 'souls'. As yet psyche is not so corrupted except perhaps in its use in such materialist disciplines as psychology and its derivatives, psychobiology, neuropsychology, psychosomatics, parapsychology, psychopharmacology. However, there are now signs that the conscious self or psyche can be referred to in 'polite' scientific discourse without evoking an outrage verging on obscenity! So I have dared to call this course of lectures – 'The Human Psyche'. Sperry (1976) has described the cautious manner in which he first introduced (Sperry, 1952) into a behaviouristically dominated psychology the notion that mental events play a significant role in decision making.

The concept of the psyche or soul appears in a clearly defined form in the Socratic dialogues of Plato, and particularly in the *Phae-*

do, where it was linked with immortality. I shall assemble the main ideas in a series of Socratic statements from the *Phaedo*, where Psyche is translated as Soul.

> 'Well now', said Socrates, 'are we not part body, part soul?' ... 'When the soul uses the instrumentality of the body for any enquiry, whether through sight or hearing or any other sense (because using the body implies using the senses), it is drawn away by the body into the realm of the variable, and loses its way'... 'When soul and body are both in the same place, nature teaches the one to serve and be subject, the other to rule and to govern'.... 'The soul is most like that which is divine, immortal, intelligible, uniform, indissoluble, and ever self-consistent and invariable, whereas body is most like that which is human, mortal, multiform, unintelligible, dissoluble and never self-consistent' ... 'in that case is it not natural for body to disintegrate rapidly, but for soul to be quite or very nearly indissoluble?'

It will be appreciated that at the time of Socrates the scientific knowledge of the body was very primitive. Hick (1976) expresses very well the significance of these dialogues in defining the conditions for the most perfect life:

> In the dialogues of Plato we see the end of the transition from primitive ideas of the survival of depleted shades or ghosts to the ardent desire for and belief in an immortality in which the highest possibilities glimpsed in this life may be fulfilled. And we see in the thought of Socrates and Plato the essential dependence of any such positive conception of immortality upon faith in an eternal realm or being, infinite in reality and value, through relationship to which men can enjoy a shared eternity. In the Dialogues the soul (*psyche*) which survives death is no longer thought of as an ethereal body but as the mind or inner self; it is no longer a quasi-physical 'thing' but the centre of consciousness and of morally responsible choice. Accordingly the most important concern in life is to attend to 'the perfecting of your soul'.... What is important is the fact that the idea of a desirable immortality, as distinguished from that of an undesired because pointless and joyless survival, arose with the emergence of individual self-consciousness and as a correlate of faith in a higher reality which was the source of value.

It will be recognized that there is a strong dualism in the Socratic–Platonic philosophy of the psyche and the body. However, the modern philosophical concept of this dualism was first formulated by Descartes in his inimitable way. I quote from his famous argument in *The Discourse on the Method* in which he attempted to discover if everything was illusory as in a dream or if there was any certainty:

> I resolved to assume that everything that ever entered into my mind was no more true than the illusions of my dreams. But immediately afterwards I noticed that whilst I thus wished to think all things false, it was absolutely essential that the 'I' who thought this should be somewhat, and remarking that this truth *'I think, therefore I am'* was so certain and so assured that all the most extravagant suppositions brought forward by the sceptics were incapable of shaking it, I came to the conclusion that I could receive it without scruple as the first principle of the Philosophy for which I was seeking. ... From that I knew that I was a substance the whole essence or nature of which is to think, and that for its existence there is no need of any place, nor does it depend on any material thing; so that this 'me', that is to say, the soul by which I am what I am, is entirely distinct from the body, and is even more easy to know than is the latter; and even if the body were not, the soul would not cease to be what it is.

This is a very strong dualism that Descartes has derived from his *Gedanken* exercise. It has been much criticized by materialists, but Badham (1976) shows that they have misunderstood what was the object of Descartes' enquiry, namely, the certainty of a 'thinking I', the 'Me-thinking'. Badham goes on to say:

> The point of this argument is that I must identify myself with my mind since I can imagine myself owning and using another body. Consider Professor Kneale's (1962) example:
> 'If I found one morning on looking into the mirror that the face I proposed to shave was unfamiliar, I should indeed be very much surprised, but I could not say that it was not my face if at the same time I had to admit that I saw with its eyes, talked with its lips, and in general had what we call inner perceptions of it' ... each of us recognizes his own (body) ... as something with which he perceives and acts.

Extremely disfiguring injuries or cancerous destruction of the face are examples for Kneale's argument. We know that personal identity is retained in quadriplegia, where the whole body and limbs are disconnected from the brain and so from the mind. Nor does transplantation of the heart or any other organ affect personal identity, which is even retained after commissurotomy as described in Lecture 1. Transplanted limbs that can be effectively used following nerve regeneration are a future possibility. There has even been successful transplantation of primate heads onto 'donor' bodies, but the acquired body and limbs cannot be used because there is no neural regeneration of tracts in the spinal cord. However, all these 'body' examples miss the point because it is the *brain–mind* problem not the body–mind problem that is at issue. The question of self identity after brain death is not answered by Descartes' *Gedanken* experiment. It is the key problem in relation to immortality and will be discussed later. Figure 1-7 gives the diagrammatic background of this discussion.

10.3 Uniqueness of the Psyche

It is not in doubt that each human person recognizes its own uniqueness, and this is accepted as the basis of social life and of law. When we enquire into the grounds for this belief, the evidence just presented eliminates an explanation in terms of the body. There are two acceptable alternatives, the brain or the psyche. Materialists must subscribe to the former, but dualist-interactionists have to regard the psyche or World 2 (cf. Fig. 1-7) as being the vehicle for the uniqueness of the experienced self. This of course is not to deny that each brain also is unique. A structure of such complexity can *never* be duplicated down to the most minute detail, and it is the minute detail that counts. This is even true when there is the genetic identity of uniovular twins. We can recall the potentiality for a virtually infinite variety of spatio-temporal patterns of the modules (Lecture 2) and their modification in the memory process (Lecture 7). But our enquiry concerns the experienced uniqueness, and this can be preserved even when there are profound changes in the brain in surgery, in cerebral injuries and in degenerations.

As a dualist-interactionist (Lecture 2), I believe that my experienced uniqueness lies not in the uniqueness of my brain, but in my psyche. It is built up from the tissue of memories of the most intimate kind from my earliest recollection (at just before 1 year) onwards to the present. I found myself as an experiencing being in Southern Australia (Victoria) in the early years of this century. I can ask: why there? and why then? These are the questions asked by Pascal. His answer was that it was God's will.

It is important to disclaim a solipsistic solution of the uniqueness of the self. Our direct experiences are of course subjective, being derived solely from our brain and self. The existences of other selves *are established* by intersubjective communication. The obvious explanation of my experienced uniqueness is that it results from my unique genetic endowment and that this is a necessary and sufficient condition. But I shall show that this solution is untenable because of its unimaginable improbability.

A provisional statistical evaluation relating to my inherited material uniqueness is derived from the process of generation of spermatozoa and ova. Each spermatozoon would have carried a genetic endowment in its 23 chromosomes that is a unique sample of the paternal 23 pairs of chromosomes. Due to the selection of 23 from the original 46 with the crossing over and recombination within each chromosome pair (Fig. 8-1), there is a shuffling of the paternal genes so that each of the 400 million or so spermatozoa in an emmission has a unique gene composition. The same is also true for the meiotic process in the maternal production of ova, each again having a unique genetic composition. The fertilized ovum that developed into one's body and brain was the material product of a chance fertilization of an ovum by one of the hundreds of millions of spermatozoa ejaculated in one sex act.

So, even in that fragment of the genetic lottery, it was an incredibly chanceful fertilization that gave me my genetic uniqueness. This simple and limited description illustrates the improbability attending success in the genetic lottery. But on top of that improbability there is piled improbability on improbability. Long ago the great biologist H.S. Jennings (1930) expressed this well, as I pointed out many years ago (Eccles, 1965, 1970), and Thorpe (1974) has also quoted Jennings with approval:

> If the existence of *me* is thus tied to the formation of a particular combination of genes, one may enter upon calculations as to the chances that I should ever have existed. What are the chances that the combination which produced me should ever have been made? If we depend on the occurrence of the exact combination of genes that as a matter of fact produced us, the odds are practically infinite against your existence or my existence.
> And what about the selves that would have come into existence if other combinations of genes had been made? If each diverse combination yields a different *self*, then there existed in the two parents the potentialities, the actual beginnings, of thousands of billions of selves, of personalities, as distinct as you and I. Each of these existed in a form as real as your existence and my existence before our component germ cells have united. Of these thousands of billions, but four or five come to fruition. What has become of the others?

And of course to go further backwards in our genetical tree makes the problem even more preposterously fantastic. An alternative estimate of improbability can be obtained from a calculation of the number of possible genetic endowments for a human being. I originally (Popper and Eccles, 1977) made the calculation on the basis of the very conservative estimate of 30,000 genes. In that case there are about $10^{10,000}$ different genetic combinations. *If my uniqueness of self is tied to the genetic uniqueness that built my brain, the odds against myself existing are $10^{10,000}$ against!* Hence I must reject this materialistic doctrine. What then determines the uniqueness of my psyche?

I have found that a frequent and superficially plausible answer to this question is the statement that the determining factor is the uniqueness of the accumulated experiences of a self throughout its lifetime. And this factor is also invoked to account for the distinctiveness of uniovular twins despite their genetic identity. It is readily agreed that my behaviour, my character, my memories and in fact the whole content of my inner conscious life are dependent on the accumulated experiences of my life; but no matter how extreme the change that can be produced by the exigencies of circumstance, I would still be the same self able to trace back my continuity in memory to my earliest remembrances at the age of 1 year or so, the same self in a quite other guise. Thus the accumulated experiences of a lifetime cannot be invoked as the determining or generating factor of the unique self, though naturally they will enormously modify all the qualities and features of that self. The situation is somewhat analogous to the Aristotelian classification into substance and accidents.

Jennings (1930) must have appreciated the fallacy of attempting to derive the uniqueness of self from the experiential history of an individual, for in searching for an explanation he develops the following remarkable speculations:

> To work this out in detail, one would apparently have to hold that the human self is an entity existing independently of genes and gene combinations, and that it merely enters at times into relations with one of the knots formed by the living web. If one particular combination or knot should not occur, it would enter into another. Then each of us might have existed with quite different characteristics from those we have – as our characteristics would indeed be different if we had lived under different environments. It could be held that there is a limited store of selves ready to play their part, that the mere occurrence of two particular cells which may or may not unite has no determining value for the existence of these selves, but merely furnishes a substratum to which for reasons unknown they may become temporarily attached. ... And what interesting corollaries might be drawn from such a doctrine, as to the further independent

existence of the selves after the dispersal of the gene combinations to which they had been attached.

Jennings is here proposing, apparently unknowingly, the creationist doctrine that each soul is a Divine creation. In this connection it is of interest to present the ideas of Hick (1972; 1976).

From a religious point of view the randomness out of which we have come is an aspect of the radical contingency of our existence. ... Our dependent status is ultimately traced by religious thought back to the dependence of the entire natural order upon the creative will of God. Thus far our emergence out of the bewildering complexity of the genetic process is not in tension with the basic theological conception of man's utter contingency as a created being.

So far I am in agreement and in fact my first Gifford series was on related themes. However Hick goes on to attack the creationist doctrine *that each soul is a new divine creation which God attaches to the growing foetus at some time between conception and birth.* He states:

If the creationist doctrine is to have any substance, so as to be worth either affirming or denying, it must entail that there are characteristics of the self which are derived neither from genetic inheritance nor from interaction with the environment. If the divinely infused soul is to have a function in the economy of human existence it must form the inner core of individuality, the unique personal essence of a human being, providing the ultimate ground of human individuation. ... However this idea of innate, but not inherited qualities, is, to say the least, highly problematic.

But surely that is just the point! It is the certainty of my inner core of unique individuality that *necessitates* the 'divine creation'. I submit that no other explanation is tenable, neither the genetic uniqueness with its impossible lottery nor the environmental differentiations, which do not *determine* one's uniqueness, but which merely modify it. Later Hick seems to misunderstand the doctrine of divine creation because he relates it to a 'special divine determination of the initial genetic makeup of each human person!' But that is *not* the doctrine. The genetic make up actualized in the newly created life is random from an almost infinite number of possibilities. *The unique individuality comes from the infused soul.*

There is a tendency in religious thinking to talk about persons or people, body plus soul, as the subjects of religious concern. However, I want to refer to the dualist-interactionist diagrams of Figs. 1-2 and 1-7 in *insisting* that World 2 (the soul or psyche) must *not* be blended in with the brain and body of World 1. There *is* an interface or frontier.

A very special problem is raised by identical twins that are completely distinct in their self-conscious experiences, alike as they appear to be to external observers. The same genetic endowments must therefore be compatible with different experiencing selves. *Evidently genetic constitution is not the necessary and sufficient condition for experienced uniqueness.* Popper (Popper and Eccles, 1977, Dialogue X) states: 'The question can be raised whether the minds of identical twins are similar, just as their bodies are similar, but *not* the question whether their minds might be identical, because their bodies, similar though they may be, cannot be identical'.

This objection is valid, but it is not in conflict with the explanation that with the identical twins there is a difference in the infused souls just as with any other foetuses, non-identical twins for example.

Though the hypothesis of the infusion of the soul into the foetus at some time between fertilization and birth accounts for the uniqueness of the experiencing self as it develops self-consciousness and human personhood, there are many perplexing problems that are not usually formulated. For example: at what stage in their evolution did hominids acquire souls? What is the fate of the soul when death occurs before it can achieve expression in self-consciousness and personhood? At what stage in the developing embryo does it acquire a soul? What happens when the foetus develops with a gross abnormality of the brain as in an anencephalic idiocy or in a double-headed monster? I prefer not to speculate in such obscure fields except to state that the infusion may be gradual, and that it cannot be immediately after fertilization else we would be in trouble with the duality of the souls of uniovular twins.

10.4 Freedom of the Will

As stated by Popper (1972) at the end of the quotation in Lecture 1, according to determinism: 'Purely physical conditions, including our physical environment make us say or accept whatever we say or accept'.

Lucas (1970) very succinctly concludes in an almost identical manner:

> Determinism ... cannot be true, because if it was, we should not take the determinists' arguments as being really arguments, but as being only conditioned reflexes. Their statements should not be regarded as really claiming to be true, but only as seeking to cause us to respond in some way desired by them. And equally, if I myself come to believe in determinism, it would be because I had been subject to certain pressures or counter-pressures, not because the arguments were valid or the conclusion true. And therefore I cannot take determinism ... seriously; for if it were true, it would destroy the possibility of its being rationally considered and recognised as such. Only a free agent can be a rational one. Reasoning, and hence truth, presupposes freedom just as much as deliberation and moral choice do.

Hick (1976) wisely states:

> The problem of free will does not hinge upon whether we are created beings (i.e. the product of forces other than ourselves), but upon whether as created beings, we have any significant freedom ... the principle that to be created is *ipso facto* to be unfree would merely serve to nullify the language of personal agency.

It is essential to recognize that *all* materialist theories of the mind (cf. Lecture 1) ultimately are reducible to determinism. It is a strange paradox that determinists

> want to be considered as rational agents arguing with other rational agents; they want their beliefs to be construed as beliefs, and subjected to rational assessment; and they want to secure the rational assent of those they argue with, not a brainwashed repetition of acquiescent patter. (Lucas, 1970).

Yet as determinists they have sawn off the 'rational' branch on which they like to think they are sitting. How long can this levitational delusion be perpetuated by wishful thinking? Superstitions die hard! We can presume that their tenacity is grounded upon their horror or fear of dualist-interactionism. This philosophy has been explained in depth in the book *The Self and Its Brain* (Popper and Eccles, 1977). There were several critical reviews – some vindictively so (cf. Mandler, 1978) – but no criticisms of substance were forthcoming and some were false. Two of Mandler's objections were based upon his complete misunderstanding of evolutionary theory![1] It is strange that in the field of brain–mind philosophy

[1] This transpired in his reply to my enquiry about the grounds for his dogmatic rejection of the language–brain development suggestions on pages 11 and 13 of our book. My reply was not published by *Science* nor answered by Mandler.

intemperate behaviour is not unusual, as for example occurred in Bunge's attack on me at the World Philosophy Conference at Düsseldorf in 1978. That became front-page headline news in several German papers! One can charitably assume that this behaviour is a cover-up for the weakness and illogicality of the determinist position. The subconscious recognition of this may be operating on them not at a rational, but at an emotional level!

Figure 1-7 is the key diagram of dualist-interactionism. It is greatly attenuated, and has to be amplified in imagination to an enormous extent. Furthermore the World 2 interaction has to be seen as in dynamic flux according to interest and attention. As already mentioned in Lecture 1 there need be no conflict with the first law of thermodynamics.

In the preceding lectures there have been many examples of actions of will across the interface of Fig. 1-7. For example attention is due to a voluntarily directed action on the modules of the liaison brain so that there is a modification of their spatio-temporal patterns of activity as revealed by the N 140 and P 400 waves in Figs. 4-5, 4-6 and 4-7. In intention there is also a modification of module activities as shown by the potentials of Figs. 4-9, 4-10, 4-11 and 4-12. It was remarkable that in Figs. 6-1, 6-2 and 6-3 many purely mental activities resulted in an increased activity of the neural machinery of the neocortex as was indicated by the increase in the circulation rate. In Lecture 7 the recall of memories is another example of action across the interface of Fig. 1-7.

In the more complex and subtle human performances described in Lectures 8 and 9 it was implied that the conscious self is able to choose one behaviour pattern rather than another. This choice freely willed carries with it a moral responsibility. It is exhibited for example in altruistic behaviour and in the control of aggressive behaviour (Lecture 8). It is exhibited also in the carrying out of all actions governed by values (Lecture 9), as for example in the scientist's dedication to the pursuit of truth. More generally it appears in the unending quest for a life consonant with the highest values – truth, goodness and beauty – that gives to life meaning and purpose.

In conclusion we can say that it is of transcendent importance to recognize that by taking thought we can influence the operation of the neural machinery of the brain. In that way we can bring about changes in the world for good or for ill. A simple metaphor is that our conscious self is in the driver's seat. Our whole life can be considered as being made up of successive patterns of choices that could lead to the feeling of fulfilment with the attendant happiness that comes to a life centred on meaning and purpose (cf. Lecture 9).

10.5 Science and Religion (cf. Peacocke, 1971)

There is a pervasive belief that religion and science are antagonistic, and that religion has been mortally defeated. This is a mistake based upon ignorance and/or prejudice. Yet atheistic materialism is the in-thing for all 'tough-minded' materialists. It is surprising that this fallacious belief has been propagated despite the fact that some of the greatest scientists of this century have recognized the necessity for a religious attitude to life and to science. In illustrating this claim in such a 'sensitive' field, I will quote directly, so that there can be no suspicion of misrepresentation based upon my wishful thinking!

My first example comes from a Chapter in Heisenberg's book *Physics and Beyond* (1971) that is entitled 'Positivism, metaphysics and religion'. Heisenberg is reconstructing a conversation that he had with Wolfgang Pauli in 1952. Pauli asks Heisenberg unexpectedly:

'Do you believe in a personal God?'

Heisenberg replies:

> 'May I rephrase your question? I myself would prefer the following formulation: Can you, or anyone else, reach the central order of things or events, whose existence seems beyond doubt, as directly as you can reach the soul of another human being? I am using the term 'soul' quite deliberately so as not to be misunderstood. If you put your question like that, I would say yes'. ...
> 'Why did you use the word 'soul' and not simply speak of another person?'
> 'Precisely because the word 'soul' refers to the central order, to the inner core of a being whose outer manifestations may be highly diverse and pass our understanding'.

Later towards the end of the conversation Heisenberg made a very strong statement:

> I agree with Max Weber that, ultimately, pragmatism bases its ethics on Calvinism, i.e., on Christianity. If we ask Western man what is good and what is evil, what is worth striving for and what has to be rejected, we shall find time and again that his answers reflect the ethical norms of Christianity even when he has long since lost all touch with Christian images and parables. If the magnetic force that has guided this particular compass – and what else was its source but the central order? – should ever become extinguished, terrible things may happen to mankind, far more terrible even than concentration camps and atom bombs. But we did not set out to look into such dark recesses; let's hope the central realm will light our way again, perhaps in quite unsuspected ways.

In 1971 I was on a panel with Heisenberg at the Nobel Prize winner's Conference at Lindau. Some of the student members of the panel and in the audience tried to push Heisenberg into making an atheistic affirmation. He resisted most strongly in a deeply moving address and very effectively silenced his disputants. It was much like that quoted above.

In his Eddington Lecture, 'The Vision of Nature', Hinshelwood (1961) linked the scientific and religious views in a passage of great feeling:

> There is a profound human instinct to seek something personal behind the processes of nature; and people are led both by intellectual and by emotional paths to the contemplation of religious questions. In so far as men seek to fuse their own personal worlds with the impersonal element in the external world they are pursuing the vision of nature; in their desire for communion with the universal and personal they pursue what I suppose through the ages they have meant by the vision of God.

In his address on the occasion of the Templeton Prize award last year, Professor Torrance who is at home both in theology and in science made a most important prophetic statement about the future relations of science and theology:

> What does seem clear to me is that the old way of thinking in terms of the couplets chance and necessity, uncertainty and determinism must now be replaced by a new way of thinking in terms of spontaneity and open-structured order, for what is revealed to us is an astonishing spontaneity in nature which yields a dynamic kind of order with an indefinite range of intelligibility which cries out for completion beyond the universe known to our natural scientific inquiries. Theologically speaking, what we are concerned with here is an understanding of the spontaneity and freedom of the created universe as grounded in the unlimited spontaneity and freedom of God the Creator. Here natural science and theological science bear closely upon one another at their boundary conditions, and what

> is needed is a more adequate doctrine of creation in which knowledge from both sides of those boundary conditions can be co-ordinated.

I agree most enthusiastically and I dare to hope that my Gifford Lectures may provide some light on this way into the future.

On the occasion of the Einstein centenary Gunter Stent (1979) has written on Einstein's religious belief. Einstein's most direct statement was in reply to a cable from Rabbi Goldstein of New York in 1929. 'Do you believe in God?'. Einstein cabled back: 'I believe in Spinoza's God who reveals himself in the harmony of all being; not in a God that concerns himself with the fate and actions of men'. Stent goes on to state:

> Since according to the monotheistic tradition, God made man in His image, man's rational affinity with the Creator provides the scientist with a reasonable expectation of eventually working out His design. That Einstein saw the role of the scientist in just this light is attested by his not infrequent references to 'Spinoza's God' in scientific discussions, of which the aphorism about God not being a dice player is merely the most famous. For instance when a theory appeared to him arbitrary or forced, Einstein might comment that 'God doesn't do anything like that', or that 'God is tricky but not malicious'. And when asked what his reaction would have been if the data gathered at the 1919 solar eclipse had failed to confirm his Theory of Relativity, Einstein replied: 'Then I would have been sorry for the dear Lord'. That almost all of the references to God were made in a humorous vein does not diminish their value as testimonials of Einstein's essentially monotheistic world outlook.

I would agree with Stent's comment. Einstein had spent his life trying to understand the design of creation, so he could feel and talk as if he were on intimate terms with God, its Creator!

I conclude this section with quotations from Max Planck whose scientific eminence rivals that of Einstein. Planck was deeply religious, as can be appreciated in these quotations from his lecture on 'Religion and Natural Science' (Planck, 1950):

> The natural scientist recognizes as immediately given nothing but the content of his sense experiences and of the measurements based on them. He starts out from this point, on a road of inductive research, to approach as best he can the supreme and eternally unattainable goal of his quest – God and His world order. Therefore, while both religion and natural science require a belief in God for their activities, to the former He is the starting point, to the latter the goal of every thought process. To the former He is the foundation, to the latter the crown of the edifice of every generalized world view. ...
> Religion and natural science do not exclude each other, as many contemporaries

> of ours would believe or fear; they mutually supplement and condition each other. The most immediate proof of the compatibility of religion and natural science, even under the most thorough critical scrutiny, is the historic fact that the very greatest natural scientists of all times – men such as Kepler, Newton, Leibniz – were permeated by a most profound religious attitude. ...
> Religion and natural science are fighting a joint battle in an incessant, never relaxing crusade against scepticism and against dogmatism, against disbelief and against superstition, and the rallying cry in this crusade has always been, and always will be: *'On to God!'*.

In these Gifford Lectures on Natural Theology I am happy to quote from the writings of these great scientists in order to refute the superstition (cf. Planck above) that there is an incessant warfare between science and religion. There are of course, notorious scientific atheists, some being in the forefront of science, for example Monod. The antireligious views of many scientists can be explained because the demands of science are so exacting that there is no time for them to consider philosophical and religious questions deeply. As a consequence an easy agnostic position is often assumed. It is my hope that these lectures of mine on Natural Theology will give scientists, philosophers and theologians an opportunity to broaden the basis of their beliefs with respect to science and religion. So strong has been the scientific attack that many ministers of religion have become demoralized; and, with the defeatist slogan 'God is dead', they turn to liberal-left activities – to good works – to replace their belief in God. There are the most disastrous consequences for our society, which now goes by the unfortunate name, 'post-Christian'. I remind you of Heisenberg's prediction, as stated above, that if the guidance of mankind by a belief in 'the central order' becomes extinguished terrible things may happen.

10.6 Death and Immortality?

We continue on from Lecture 6 where there was a brief description of the deep unconsciousness of coma that may continue for months and even years with no hope of recovery. It is essential to realize that death is now correctly identified with *brain death*. With modern survival techniques a body can be kept 'alive' for months despite the death of the brain, which can be recognized with assurance by the continued 'flatness' of the EEG (cf. Fig. 6-4). This tragic situation of brain death may arise because of a too long failure of circulation in a cardiac failure or from an episode of anoxia resulting from a failure of respiration, e.g. in drowning. However, in many cases of vigil coma the neocortex is not dead, but merely deeply depressed because of the failure of the reticular activating system (cf. Lecture 6).

As already stated, on all materialist theories of the mind there can be no consciousness of any kind after brain death. Immortality is a non-problem. But with dualist-interactionism it can be recognized from the standard diagram (Fig. 1-7) that death of the brain need not result in the destruction of the central component of World 2. All that can be inferred is that World 2 ceases to have any relationship with the brain and hence will lack all sensory information and all motor expression. There is no question of a continued shadowy or ghost-like existence in some relationship with the material world, as is claimed in some spiritualist beliefs. What then can we say?

Belief in some life after death came very early to mankind, as is indicated by the ceremonial burial customs of Neanderthal man. However in our earliest records of beliefs about life after death it was most unpleasant. This can be seen in the Epic of Gilgamesh or in the Homeric poems, or in the Hebrew belief about Sheol. Hick (1976) points out that the misery and unhappiness believed to attend the life hereafter very effectively disposes of the explanation that such beliefs arose from wish-fulfilment!

The idea of a more attractive after-life is a special feature of the Socratic dialogues, being derived from the Orphic mysteries, as has already been noted in the quotations from the *Phaedo* and from Hick (1976). There was a particularly clear affirmation of immortality by Socrates in the *Phaedo* just before his death:

> 'If the soul is immortal, it demands our care not only for that part of the time which we call life, but for all time; ... since the soul is clearly immortal, it can have no escape or security from evil except by becoming as good and wise as it possibly can. For it takes nothing with it to the next world except its education and training'.
> ... Said Crito. 'But how shall we bury you?' 'Any way you like,' replied Socrates, 'that is if you can catch me and I don't slip through your fingers'. He laughed gently as he spoke, and turning to us went on: 'I can't persuade Crito that I am this Socrates here who is talking to you now and marshalling all the arguments; he thinks that I am the one whom he will see presently lying dead; and he asks how he is to bury me! As for my long and elaborate explanation that when I have drunk the poison I shall remain

with you no longer, but depart to a state of heavenly happiness, this attempt to console both you and myself seems to be wasted on him'.

After the poignant simplicity of Socrates' messages before death, it is quite an experience to contemplate the many kinds of immortality that have been the subject of speculation. The idea of immortality has been sullied over and even made repugnant by the many attempts from earliest religions to give an account that was based on the ideologies of the time. Thus today intellectuals are put off by these archaic attempts to describe and depict life after bodily death. I am put off by them too. It is not my task in these lectures to pursue such theological complexities; but I shall give some references to books that I have read. The stories of man's thoughts on the ultimate human destiny in death have been collected in two scholarly books by Choron (1963, 1964). Theological and philosophical disputations are given in books by von Hügel (1912), Fosdick (1918), Baillie (1934), Dixon (1937), D'Arcy (1942), Pelikan (1962); and more recently in the books that I have already referred to – Hick (1976), Lewis (1978), and Badham (1976).

In these recent books there has been much discussion on the resurrection of the body, particularly by Hick, because it is commonly assumed that the soul can only function in relation to some kind of body. In this connection there is discussion of the strange concept of 'astral bodies', which I find quite unacceptable. I would suggest that it is not valuable to speculate on this 'body'–soul problem after death. It is perplexing enough during life! Self-recognition and communication may be possible for the psyche in ways beyond our imagination.

A more interesting and meaningful disputation concerns the recognition of self after death. We normally have the body and brain to assure us of our identity, but, with departure of the psyche from the body and brain in death, none of these landmarks is available to it. All of the detailed memory must be lost. If we refer again to Fig. 1-7, memory is also shown located in World 2. I would suggest that this is a more general memory related to our self-identity, our emotional life, our personal life and to our ideals as enshrined in the values. All of this should be sufficient for self-identity. Reference should be made to the discussion on the creation of the psyche by infusion into the developing embryo. *This divinely created psyche should be central to all considerations of immortality and of self-recognition* (cf. Lewis, 1978).

Another topic of speculation concerns the avoidance of tedium in the endless time of eternity. Lemberg (1980) makes an important comment:

> I believe that it is generally neglected, yet it is most essential to differentiate between the idea of life after death and the concept of eternity. Most people, whether scientific, commonsense or religious, do not make this differentiation. For them they are synonymous. Yet this is a profound misconception of 'eternity'. About 250 A.D. Plotinus stressed that Eternity predates time, and was latent in the Eternal Being before the creation of the universe and of time.

In this context we can remember the description of the Big Bang in Lecture 2 of my first Gifford series (Eccles, 1979a). Hick (1976) expresses very subtle ideas on the transformation of the self in the life hereafter:

> The distinction between the self as ego and the self as person suggests that as the human individual becomes perfected he becomes more and more a person and less and less an ego. Since personality is essentially outward-looking, as a relationship to other persons, whilst the ego forms a boundary limiting true personal life, the perfected individual will have become a personality without egoity, a living consciousness which is transparent to the other consciousnesses in relation to which it lives in a full community of love. Thus we have the picture of a plurality of personal centres without separate peripheries. They will have ceased to be mutually exclusive and will have become mutually inclusive and open to one another in a richly complex shared consciousness. The barrier between their common unconscious life and their individual consciousness will have disappeared, so that they experience an intimacy of personal community which we can at present barely imagine.

This formulation by Hick may be correlated with the religious belief in the communion of souls. Need we speculate further? Just before he died the great biochemist Lemberg (1980) wrote very movingly on his impending death:

> I can only speak for myself but I am satisfied to leave judgement on what my life here on earth has been in the hands of the Eternal, recognizing that by the grace of God I have not been so wronged in my life that in spite of suffering the fate of a refugee I shall have to expect recompense in another life. Whether there will be a life after death for any man or woman I do not know and cannot know.

The famous author Freya Stark (1948) wrote most movingly on ageing, death and the hope of a life hereafter:

> Our private grasp lessens, and leaves us heirs to infinite loves in a common world where every joy is a part of one's personal joy. With a loosening hold returning towards acceptance, we prepare in the anteroom for a darkness where even this last personal flicker fades, and what happens will be in the Giver's hand alone.

10.7 Epilogue

The theme of my previous Gifford lectures was 'The Human Mystery'. Our life here on this earth and cosmos is beyond our understanding in respect of the Great Questions. We have to be open to some deep dramatic significance in this earthly life of ours that may be revealed after the transformation of death. We can ask: What does this life mean? We find ourselves here in this wonderfully rich and vivid conscious experience and it goes on through life; but is that the end? This self-conscious mind of ours has this mysterious relationship with the brain and as a consequence achieves experiences of human love and friendship, of the wonderful natural beauties, and of the intellectual excitement and joy given by appreciation and understanding of our cultural heritages. Is this present life all to finish in death or can we have hope that there will be further meaning to be discovered? In the context of Natural Theology I can only say that there is complete oblivion about the future; but we came from oblivion. Is it that this life of ours is simply an episode of consciousness between two oblivions, or is there some further transcendent experience of which we can know nothing until it comes?

Man has lost his way ideologically in this age. It is what has been called the predicament of mankind. I think that science has gone too far in breaking down man's belief in his spiritual greatness, as exemplified in the magnificent achievements in World 3, and has given him the belief that he is merely an insignificant animal that has arisen by chance and necessity in an insignificant planet lost in the great cosmic immensity. This is the message given to us in the quotations from Monod in Lecture 9. I think the principal trouble with mankind today is that the intellectual leaders are too arrogant in their self-sufficiency. We must realize the great unknowns in the material makeup and operation of our brains, in the relationship of brain to mind and in our creative imagination. When we think of these unknowns as well as the unknown of how we come to be in the first place, we should be much more humble. The unimaginable future that could be ours would be the fulfillment of this our present life, and we should be prepared to accept its possibility as the greatest gift. In the acceptance of this wonderful gift of life and of death, we have to be prepared not for the inevitability of some other existence, but we can hope for the possibility of it.

This is the message we would get from what Penfield (1975) and Thorpe (1962) have written; and I myself have also the strong belief that we have to be open to the future. This whole cosmos is not just running on and running down for no meaning. In the context of Natural Theology I come to the belief that we are creatures with some supernatural meaning that is as yet ill defined. We cannot think more than that we are all part of some great design, which was the theme of my first Gifford series (Eccles, 1979a). Each of us can have the belief of acting in some unimaginable supernatural drama. We should give all we can in order to play our part. Then we wait with serenity and joy for the future revelations of whatever is in store after death.

It is not my task in the context of Natural Theology to say more or to make special reference to the highest religious beliefs as enshrined in Christianity. But I would claim that in its entirety my deliberations on Natural Theology in these two series of Gifford Lectures are not antithetic to Christianity, in which I am a believer.

References

[Numbers in square brackets at end of each entry indicate the pages on which it is cited]

Adrian, E.D.: The physical background of perception. Oxford: Clarendon 1946 [82, 83, 87, 88]

Adrian, E.D., Matthews, B.H.C.: The Berger rhythm: Potential changes from the occipital lobes of man. Brain *57*, 355–384 (1934) [80, 81, 82]

Ajmone-Marsan, C.: Acute effects of topical epileptogenic agents. In: Basic mechanisms of the epilepsies. Jasper, H.H., Ward, A.A., Pope, A. (eds.), pp. 299–319. Boston: Little, Brown & Co. 1969 [159]

Albe-Fessard, D., Besson, J.M.: Convergent thalamic and cortical projections. The non-specific system. In: Somatosensory system. Iggo, A. (ed.), pp. 489–560. Berlin, Heidelberg, New York: Springer 1973 [127]

Allen, G.I., Tsukahara, N.: Cerebrocerebellar communication systems. Physiol. Rev. *54*, 957–1006 (1974) [105]

Amsterdam, B.: Mirror self-image reactions before the age of two. Dev. Psychobiol. *5*, 297–305 (1972) [3]

Andersen, P., Andersson, S.A.: Physiological basis of the alpha rhythm. New York: Appleton-Century-Crofts 1968 [84, 85]

Andersson, S.A., Keller, O., Vyklický, L.: Cortical activity evoked from tooth pulp afferents. Brain Res. *50*, 473–475 (1973) [127]

Andy, O.J., Stephan, H.: The septum in the human brain. J. Comp. Neurol. *133*, 383–409 (1968) [112]

Ardrey, R.: African genesis. New York: Atheneum 1961 [209]

Ardrey, R.: The territorial imperative. New York: Atheneum 1966 [209]

Armstrong, D.M.: A materialist theory of the mind. London: Routledge & Kegan Paul 1968 [3]

Armstrong, D.M.: Between matter and mind. London: Times Literary Suppl. Feb. 17, 1978 [174]

Asanuma, H., Rosén, I.: Topographical organization of cortical efferent zones projecting to distal forelimb muscles in the monkey. Exp. Brain Res. *14*, 243–256 (1972) [31]

Attneave, F.: Multistability in perception. In: Image, object and illusion. Held, R. (ed.), pp. 90–99. San Francisco: Freeman 1974 [67, 68, 71]

Austin, G., Hayward, W., Rouhe, S.: A note on the problem of conscious man and cerebral disconnection by hemispherectomy. In: Symposium on cerebral dysconnection. Smith, L. (ed.). Springfield, Ill.: Thomas 1972 [163]

Badham, P.: Christian beliefs about life after death. London: Society for Promotion of Christian Knowledge 1976 [234, 236, 249]
Baier, K.: What is value? An analysis of the concept. In: Values and the Future. Baier, K., Rescher, N. (eds.), pp. 35–67. New York: Free Press 1969 [219]
Baillie, J.: And the life everlasting. London: Oxford University Press 1934 [249]
Barlow, H.B.: Single units and sensation: A neuron doctrine for perceptual psychology? Perception *1*, 371–394 (1972) [17, 25]
Barlow, H.B., Blakemore, C., Pettigrew, J.D.: The neural mechanism of binocular depth discrimination. J. Physiol. (London) *193*, 327–342 (1967) [74]
Becker, W., Iwase, K., Jürgens, R., Kornhuber, H.H.: Brain potentials preceding slow and rapid hand movements. In: The responsive brain. McCallum, Knott (eds.), pp. 99–102. Bristol: Wright 1976 [103]
Beloff, J.: The existence of mind. London: MacGibbon & Kee 1962 [3]
Beloff, J.: Mind–body interactionism in the light of the parapsychological evidence. Theoria to Theory *10*, 125–137 (1976) [18, 97]
Berger, H.: Über das Elektrenkephalogramm des Menschen. Arch. Psychiatr. *87*, 527–570 (1929) [80]
Bert, J., Balzamo, E., Vuillon-Cacciuttolo, G., Naquet, R.: Patterns of sleep in primates: phylogenetic, psychophysiologic and pathologic effects. In: Cerebral correlates of conscious experience. Buser, P., Rougeul-Buser, A. (eds.), pp. 233–244. Amsterdam: North Holland 1978 [154]
Bertram, B.C.R.: The social system of lions. Sci. Am. *232/5*, 54–65 (1975) [202]
Birch, C.: Chance, necessity and purpose. In: Studies in the philosophy of biology. Ayala, F.J., Dobzhansky, T. (eds.), pp. 225–239. London: Macmillan 1974 [20, 25]
Bishop, P.O.: Cortical beginning of visual form and binocular depth perception. In: Neurosciences research study program. Schmitt, F.O. (ed.), pp. 471–484. New York: Rockefeller University Press 1970 [74]
Blakemore, C.: The baffled brain. In: Illusion in nature and art. Gregory, R.L., Gombrich, E.H. (eds.), pp. 9–47. London: Duckworth 1973 [63, 75]
Blakemore, C.: Mechanics of the mind. Cambridge, London: Cambridge University Press 1977 [3, 17]
Bliss, T.V.P., Lømo, T.: Long-lasting potentiation of synaptic transmission in the dentate area of the anaesthetized rabbit following stimulation of the perforant path. J. Physiol. (London) *232*, 331–356 (1973) [177]
Bogen, J.E.: The other side of the brain II: An appositional mind. Bull. Los Angeles Neurol. Soc. *34*, 135–162 (1969) [14]
Bowsher, D.: An anatomophysiological basis of somatosensory discrimination. Int. Rev. Neurobiol. *8*, 35–75 (1965) [127]
Bremer, F.: L'activité cérébrale au cours du sommeil et de la narcose. Contribution à l'étude du méchanisme du sommeil. Bull. Acad. R. Med. Belg. *2*, 68–86 (1937) [145]
Bremer, F.: Some problems in neurophysiology. London: Athlone 1953 [145]
Bremer, F.: The neurophysiological problem of sleep. In: Brain mechanisms and consciousness. Adrian, E.D., Jasper, H.H., Bremer, F. (eds.), pp. 137–162. Springfield, Ill.: Thomas 1954 [145, 153]
Bremer, F.: Cerebral hypnogenic centers. Ann. Neurol. *2*, 1–6 (1977) [154]
Brodal, A.: Neurological anatomy. In relation to clinical medicine. London: Oxford University Press 1969 [111]

Brodmann, K.: Vergleichende Lokalisationslehre der Großhirnrinde. Leipzig: Barth 1909 [31, 55]

Brodmann, K.: Neue Ergebnisse über die vergleichende histologische Lokalisation der Großhirnrinde. Anat. Anz. *41*, 157–216 (1912) [31]

Bronowski, J.: The ascent of man. London: British Broadcasting Corporation 1973 [189]

Bunge, M.: Emergence and the mind. Neuroscience *2*, 501–509 (1977) [18, 23, 24]

Bunge, M., Llinás, R.: The mind–body problem in the light of neuroscience. 16th World Congress of Philosophy, pp. 131–133. Düsseldorf: 1978 [23, 24]

Burton, H., Jones, E.G.: The posterior thalamic region and its cortical projection in new world and old world monkeys. J. Comp. Neurol. *168*, 249–301 (1976) [126]

Campbell, D.T.: On the conflicts between biological and social evolution and between psychology and moral tradition. Am. Psychol. Vol. 30, 1103–1126 (1975) [199, 201, 206]

Carreras, M., Andersson, S.A.: Functional properties of neurons of the anterior ecto-sylvian gyrus of the cat. J. Neurophysiol. *26*, 100–126 (1963) [125]

Choron, J.: Death and western thought. London: Ballica-Macmillan 1963 [249]

Choron, J.: Modern man and mortality. New York: MacMillan 1964 [249]

Clemente, C.D., Chase, M.H.: Neurological substrates of aggressive behaviour. Annu. Rev. Physiol. *35*, 329–356 (1973) [115]

Cournand, A., Meyer, M.: The scientist's code. Minerva: A Review of Science, Learning and Policy *14*, 79–96 (1976) [223]

Creutzfeldt, O.D.: Neuronal mechanisms underlying the EEG. In: Brain mechanisms of the epilepsies. Jasper, H.H., Ward, A.A., Pope, A. (eds.), pp. 397–410. Boston: Little, Brown & Co. 1969 [159]

Creutzfeldt, O.D.: Neurophysiological correlates of different functional states of the brain. In: Brain work: The coupling of function metabolism and blood flow in the brain. Ingvar, D.H., Lassen, N.A. (eds.), pp. 21–46. Copenhagen: Munksgaard 1975 [140, 160, 161, 163]

Creutzfeldt, O.D.: Generality of the functional structure of the neocortex. Naturwissenschaften *64*, 507–517 (1977) [30]

Creutzfeldt, O.D., Jung, R.: Neuronal discharge in the cat's motor cortex during sleep and arousal. In: The nature of sleep. Wolstenholme, G.E.W., O'Connor, M. (eds.), pp. 129–170. London: Churchill 1961 [148, 150]

Creutzfeldt, O.D., Rager, G.: Brain mechanisms and the phenomenology of conscious experience. In: Cerebral correlates of conscious experience. Buser, P.A., Rougeul-Buser, A. (eds.), pp. 311–318. Amsterdam: North Holland 1978 [26]

D'Arcy, M.C.: Death and life. London: Longmans Green & Co. 1942 [249]

Dart, R.A.: The predatory transition from ape to man. Anthropol. Linguistic Rev. *1*, 201–213 (1953) [209]

Dart, R.: Adventures with the missing link. New York: Harper & Row 1959 [209]

Dawkins, R.: The selfish gene. Oxford: Oxford University Press 1976 [166, 194, 195, 198, 200]

Deecke, L., Kornhuber, H.H.: An electrical sign of participation of the mesial 'supplementary' motor cortex in human voluntary finger movement. Brain Res. *159*, 473–476 (1978) [105]

Deecke, L., Scheid, P., Kornhuber, H.H.: Distribution of readiness potential, premotion positivity, and motor potential of the human cerebral cortex preceding voluntary finger movements. Exp. Brain Res. *7*, 158–168 (1969) [102]

Deecke, L., Grözinger, B., Kornhuber, H.H.: Voluntary finger movements in man. Cerebral potentials and theory. Biol. Cybern. *23*, 99–119 (1976) [100]

Deecke, L., Englitz, H.G., Kornhuber, H.H., Schmitt, G.: Cerebral potentials preceding voluntary movement in patients with bilateral or unilateral Parkinson akinesia. In: Progress in clinical neurophysiology, Vol. 1: Attention, voluntary contraction and event-related cerebral potentials. Desmedt, J.E. (ed.), pp. 151–163. Basel: Karger 1977 [104]

Delgado, J.M.R.: Physical control of the mind. New York: Harper & Row 1969 [115, 116]

Deregowski, J.B.: Illusion and culture. In: Illusion in nature and art. Gregory, R.L., Gombrich, E.H. (eds.), pp. 161–191. London: Duckworth 1973 [77]

Deregowski, J.B.: Pictorial perception and culture. In: Image, object and illusion. Held, R. (ed.), pp. 79–85. San Francisco: Freeman 1974 [77]

Descartes, R.: Philosophical works. Cambridge: Cambridge University Press 1931 [235]

Desmedt, J.E.: Active touch exploration of extrapersonal space elicits specific electrogenesis in the right cerebral hemisphere of intact right-handed man. Proc. Natl. Acad. Sci. USA *74*, 4037–4040 (1977) [93, 95]

Desmedt, J.E., Robertson, D.: Differential enhancement of early and late components of the cerebral somatosensory evoked potentials during forced-paced cognitive tasks in man. J. Physiol. (London) *271*, 761–782 (1977a) [46, 89, 91, 93, 97]

Desmedt, J.E., Robertson, D.: Search for right hemisphere asymmetries in event-related potentials to somatosensory cueing signals. In: Language and hemispheric specialization in man: cerebral event-related potentials. Desmedt, J.E. (ed.), pp. 172–187. Basel: Karger 1977b [46, 89, 90, 91, 93, 94, 97]

De Witt, L.W.: Consciousness, mind and self: the implications of the split-brain studies. Br. J. Phil. Sci. *26*, 41–47 (1975) [13, 14]

Dixon, W.M.: The human situation: problems of life and destiny. London: Arnold 1937 [249]

Dobzhansky, T.: The biology of ultimate concern. New York: New American Library 1967 [206]

Doty, R.W.: Consciousness from neurons. Acta Neurobiol. Exp. (Warsz.) *35*, 791–804 (1975) [17, 25]

Dubos, R.: So human an animal. New York: Scribner's Sons 1968 [228]

Eccles, J.C.: Interpretation of action potentials evoked in the cerebral cortex. Electroencephalogr. Clin. Neurophysiol. *3*, 449–464 (1951a) [86]

Eccles, J.C.: Hypotheses relating to the brain–mind problem. Nature *168*, 53–57 (1951b) [22]

Eccles, J.C.: The neurophysiological basis of mind: The principles of neurophysiology. Oxford: Clarendon 1953 [22]

Eccles, J.C.: The brain and the unity of conscious experience (Eddington Lecture). London: Cambridge University Press 1965 [238]

Eccles, J.C.: Facing reality. Berlin, Heidelberg, New York: Springer 1970 [238]

Eccles, J.C.: An instruction-selection theory of learning in the cerebellar cortex. Brain Res. *127*, 327–352 (1977) [180]

Eccles, J.C.: An instruction-selection hypothesis of cerebral learning. In: Cerebral correlates of conscious experience. Buser, P.A., Rougeul-Buser, A. (eds.), pp. 155–175. Amsterdam: North Holland 1978 [176, 177, 178, 180, 181, 182, 184, 186, 187]

Eccles, J.C.: The human mystery. Berlin, Heidelberg, New York: Springer 1979a [2, 3, 14, 23, 33, 47, 77, 102, 112, 162, 166, 175, 176, 177, 178, 181, 184, 187, 195, 203, 210, 216, 223, 229, 250, 252]

Eccles, J.C.: Synaptic plasticity. Naturwissenschaften *66*, 147–153 (1979b) [177]

Eccles, J.C., Gibson, W.C.: Sherrington, his life and thought. Berlin, Heidelberg, New York: Springer 1979 [22]

Eddington, A.S.: The philosophy of physical science. London: Cambridge University Press 1939 [223, 224]

Edelman, G.M.: Group selection and phasic reentrant signaling: A theory of higher brain function. In: The mindful brain. Schmitt, F.O., pp. 51–100. Cambridge, Mass.: MIT Press 1978 [17, 49]

Egger, M.D., Flynn, J.P.: Further studies on the effects of amygdaloid stimulation and ablation on hypothalamically elicited attack behaviour in cats. In: Structure and function of the limbic system. Adey, W.R., Tokizane, T. (eds.), pp. 165–182. Amsterdam: Elsevier 1967 [114]

Evarts, E.V.: Temporal patterns of discharge of pyramidal tract neurons during sleep and waking in the monkey. J. Neurophysiol. *27*, 152–171 (1964) [148, 149, 151, 163]

Evarts, E.V.: Unit activity in sleep and wakefulness. In: The neurosciences. Quarton, G.C., Melnechuk, T., Schmitt, F.O. (eds.), pp. 545–556. New York: Rockefeller University Press 1967 [33, 148, 151, 152, 163]

Feigl, H.: The "mental" and the "physical". Minneapolis, Minn.: University of Minnesota Press 1967 [18]

Fernandez de Molina, A., Hunsperger, R.W.: Central representation of affective reactions in forebrain and brain stem: electrical stimulation of amygdala, stria terminalis, and adjacent structures. J. Physiol. (London) *145*, 251–265 (1959) [114]

Fields, H.L., Basbaum, A.I.: Brainstem control of spinal pain-transmission neurons. Annu. Rev. Physiol. *40*, 217–248 (1978) [128]

Fifková, E., Van Harreveld, A.: Long-lasting morphological changes in dendritic spines of dentate granular cells following stimulation of the entorhinal area. J. Neurocytol. *6*, 211–230 (1977) [177]

Fosdick, H.E.: The assurance of immortality. London: Clarke 1918 [249]

Frankl, V.E.: Reductionism and nihilism. In: Beyond reductionism. Koestler, A., Smythies, J.R. (eds.), pp. 396–416. London: Hutchinson 1969 [229]

Fuxe, K.: The neural circuitry of sleep – Presentation. In: Perspectives in the brain sciences, Vol. 1: The sleeping brain. Chase, M.H. (ed.), pp. 122–127. Los Angeles: Brain Research Institute University of California 1972 [154]

Gallup, G.G.: Self-recognition in primates. Am. Psychol. *32*, 329–338 (1977) [2]

Gazzaniga, M.S.: The split brain in man. In: Physiological psychology. Thompson, R.F. (ed.) San Francisco: Freeman 1971, pp. 118–123 [14]

Gibson, J.J.: The senses considered as perceptual systems. London: Allen & Unwin 1966 [4]

Goldman, P.S., Nauta, W.J.H.: Columnar distribution of cortico-cortical fibers in the frontal association, limbic and motor cortex of the developing rhesus monkey. Brain Res. *122*, 393–413 (1977) [34, 35, 47, 56, 58, 96, 178, 181–184, 187]

Gombrich, E.H.: Art and illusion. London: Phaidon 1960 [59, 75]

Gombrich, E.H.: The visual image. Sci. Am. *227*, No. 3, 82–96 (1972) [67]

Gombrich, E.H.: Illusion and art. In: Illusion in nature and art. Gregory, R.L., Gombrich, E.H. (eds.), pp. 193–243. London: Duckworth 1973 [67, 68, 71, 75]

Gordon, H.W., Bogen, J.E., Sperry, R.W.: Absence of deconnexion syndrome in two patients with partial section of the neocommissures. Brain *94*, 327–336 (1971) [11]

Granit, R.: The purposive brain. Cambridge, Mass.: MIT Press 1977 [49, 227]

Graybiel, A.M.: The thalamo-cortical projection of the so-called posterior nuclear group: A study with anterograde degeneration methods in the cat. Brain Res. *49*, 229–244 (1973) [123, 127, 130]

Graybiel, A.M.: Studies on the anatomical organization of posterior association cortex. In: The neurosciences. Schmitt, F.O., Worden, F.G. (eds.), pp. 205–214. Cambridge, Mass.: MIT Press 1974 [123, 126]

Gregory, R.L.: Perceptual illusions and brain models. Proc. R. Soc. Lond. B *171*, 279–296 (1968) [62, 64, 65]

Gregory, R.L.: Eye and brain. London: World University Library 1972 [63, 64, 65, 66, 67, 69, 70, 76]

Gregory, R.L.: The confounded eye. In: Illusion in nature and art. Gregory, R.L., Gombrich, E.H. (eds.), pp. 49–95. London: Duckworth 1973 [60, 65, 66, 67, 68, 77]

Gregory, R.L.: Visual illusions. In: Image, object and illusion. Held, R. (ed.), pp. 48–58. San Francisco: Freeman 1974 [67, 68]

Griffin, D.R.: The question of animal awareness. New York: Rockefeller University Press 1976 [2]

Grünewald-Zuberbier, E., Grünewald, G., Jung, R.: Slow potentials of the human precentral and parietal cortex during goal-directed movements (Zielbewegungspotentiale). J. Physiol. (London) *284*, 181–182P (1978) [99, 101]

Hardy, J.D., Stolwijk, J.A.J.: Tissue temperature and thermal pain. In: Touch, heat and pain. Ciba Symposium. De Reuck, A.U.S., Knight, J. (eds.), pp. 27–50. London: Churchill 1966 [123]

Hassler, R.: Die zentralen Systeme des Schmerzes. Acta Neurochir. (Wien) *8*, 353–423 (1960) [127]

Hassler, R.: Central interactions of the systems of rapidly and slowly conducted pain. In: Advances in neurosurgery. Penzholz, H., Brock, M., Hamer, J., Klinger, M., Spoerri, O. (eds.), Vol. 3. pp. 143–150. Berlin, Heidelberg, New York: Springer 1975 [124, 127]

Hassler, R.: Interaction of reticular activating system for vigilance and the truncothalamic and pallidal systems for directing awareness and attention under striatal control. In: Cerebral correlates of conscious experience. Buser, P.A., Rougeul-Buser, A. (eds.), pp. 111–129. Amsterdam: North Holland 1978 [131, 160]

Hawkes, J.: Prehistory in history of mankind. Cultural and scientific development. Vol. 1, Part. 1. Unesco. London: New English Library 1965 [206]

Hayek, F.A.: The three souces of human values. London: The London School of Economics and Political Science 1978 [211, 216, 217]

Heath, R.G.: Studies in schizophrenia. A multidisciplinary approach to mind–brain relationship. Cambridge, Mass.: Harvard University Press 1954 [116]

Heath, R.G.: Electrical self-stimulation of the brain in man. Am. J. Psychiat. *120*, 571–577 (1963) [116]

Hecaen, H.: Brain mechanisms suggested by studies of parietal lobes. In: Brain mechanisms underlying speech and language. Millikan, C.H., Darley, F.L. (eds.), pp. 146–166. New York, London: Grune & Stratton 1967 [130]

Heisenberg, H.: Physics and beyond, encounters and conversations. London: Allen & Unwin 1971 [223, 224, 244, 245]

Held, R.: Object and effigy. In: Structure in art and in science. Kepes, G. (ed.), pp. 42–54. New York: Braziller 1965 [52]

Hess, W.R.: Beiträge zur Physiologie des Hirnstammes. I. Die Methodik der lokalisierten Reizung und Ausschaltung subkortikaler Hirnabschnitte. Leipzig: Thieme 1932 [113]

Hess, W.R., Brügger, M.: Das subkorticale Zentrum der affektiven Abwehrreaktion. Helvet. Physiol. Acta *1*, 33–52 (1943) [114]

Hick, J.: Biology and the soul. Cambridge: Cambridge University Press 1972 [240]

Hick, J.: Death and eternal Life. London: Collins 1976 [234, 235, 240, 242, 248, 249, 250]

Hinshelwood, C.: The vision of nature. Cambridge: Cambridge University Press 1961 [223, 245]

Hubel, D.H., Wiesel, T.N.: Shape and arrangement of columns in the cat's striate cortex. J. Physiol. (London) *165*, 559–568 (1963) [31]

Hubel, D.H., Wiesel, T.N.: Receptive fields and functional architecture in two nonstriate visual areas (18 and 19) of the cat. J. Neurophysiol. *28*, 229–289 (1965) [57]

Hubel, D.H., Wiesel, T.N.: Stereoscopic vision in macaque monkey. Nature 225, 41–42 (1970) [74]

Hubel, D.H., Wiesel, T.N.: Laminar and columnar distribution of geniculo-cortical fibers in the Macaque monkey. J. Comp. Neurol. *146*, 421–450 (1972) [31]

Hubel, D.H., Wiesel, T.N.: Sequence regularity and geometry of orientation columns in the monkey striate cortex. J. Comp. Neurol. *158*, 267–294 (1974) [31]

Hubel, D.H., Wiesel, T.N.: Functional architecture of Macaque monkey visual cortex. Proc. R. Soc. Lond. B *198*, 1–59 (1977) [31]

Hügel, F. von: Eternal life. A study of its implications and applications. Edinburgh: Clark 1913 [249]

Hughes, J.: Isolation of an endogenous compound from the brain with pharmacological properties similar to morphine. Brain Res. *88*, 295–308 (1975) [129]

Iggo, A.: A single unit analysis of cutaneous receptors with C afferent fibres. In: Pain and itch: Nervous mechanisms. Ciba Foundation. Wolstenholme, G.E.W., O'Connor, M. (eds.), pp. 41–59. London: Churchill 1959 [122]

Iggo, A.: Activation of cutaneous nociceptors and their actions on dorsal horn neurons. Adv. Neurol. *4*, 1–9 (1974) [122, 123]

Ingvar, D.H.: Patterns of brain activity revealed by measurements of regional cerebral blood flow. In: Brain work. Ingvar, D.H., Lassen, N.A. (eds.), pp. 397–413. Copenhagen: Munksgaard 1975a [87, 135–139]

Ingvar, D.H.: Brain work in presenile dementia and in chronic schizophrenia. In: Brain work. Ingvar, D.H., Lassen, N.A. (eds.), pp. 478–492. Copenhagen: Munksgaard 1975b [135, 136]

Inubushi, S., Kobayashi, T., Oshima, T., Torii, S.: Intracellular recordings from the motor cortex during EEG arousal in unanaesthetized brain preparations of the cat. Jpn. J. Physiol. *28*, 669–688 (1978a) [143, 163]

Inubushi, S., Kabayashi, T., Oshima, T., Torii, S.: An intracellular analysis of EEG arousal in cat motor cortex. Jpn. J. Physiol. *28*, 689–708 (1978b) [143, 163]

Isaac, G.: The food-sharing behaviour of protohuman hominids. Sci. Am. *238/4*, 90–108 (1978) [203, 215]

Isaacson, R.L.: The limbic system. New York: Plenum 1974 [115]

James, R.: Mind over matter. Higher Education Review, July 14th (1978) [174]

Jasper, H.H.: Pathophysiological studies of brain mechanisms in different states of consciousness. In: Brain and conscious experience. Eccles, J.C. (ed.), pp. 256–282. Berlin, Heidelberg, New York: Springer 1966 [87, 141]

Jasper, H.H.: Methods of propagation: Extracellular studies. In: Brain mechanisms of the epilepsies. Jasper, H.H., Ward, A.A., Pope, A. (eds.), pp. 421–440. Boston: Little, Brown 1969 [158]

Jaynes, J.: The origin of consciousness in the breakdown of the bicameral mind. Boston: Houghton Mifflin 1976 [4, 156, 216]

Jennings, H.S.: The biological basis of human nature. New York: Norton 1930 [238, 239, 240]

Jones, E.G., Leavitt, R.Y.: Demonstration of thalamo-cortical connectivity in the cat somatosensory system by retrograde axonal transport of horseradish peroxidase. Brain Res. *63*, 414–418 (1973) [127]

Jones, E.G., Powell, T.P.S.: Connections of the somatic sensory cortex of the rhesus monkey. I. Ipsilateral cortical connections. Brain *92*, 477–502 (1969) [123, 125]

Jones, E.G., Powell, T.P.S.: Electron microscopy of the somatic sensory cortex of the cat. II. The fine structure of layers I and II. Philos. Trans. R. Soc. Lond. B *257*, 13–21 (1970a) [178, 180, 186]

Jones, E.G., Powell, T.P.S.: An anatomical study of converging sensory pathways within the cerebral cortex of the monkey. Brain *93*, 793–820 (1970b) [54, 92, 123, 129]

Jones, E.G., Powell, T.P.S.: Anatomical organization of the somatosensory cortex. In: Handbook of sensory physiology. Iggo, A. (ed.), Vol. II, pp. 579–620. Berlin, Heidelberg, New York: Springer 1973 [53]

Jouvet, M.: Some monoaminergic mechanisms controlling sleep and waking. In: Brain and human behaviour. Karczmar, A.G., Eccles, J.C. (eds.), pp. 131–161. Berlin, Heidelberg, New York: Springer 1972 [153]

Jouvet, M.: Monoamine regulation of the sleep-waking cycle in the cat. In: The neurosciences: Third study program. Schmitt, F.O., Worden, F.G. (eds.), pp. 499–508. Boston, Mass.: MIT Press 1974 [154]

Julesz, B.: Foundations of cyclopean perception. Chicago: University of Chicago Press 1971 [72, 73, 74, 77]

Julesz, B.: Texture and visual perception. In: Image, object and illusion. Held, R. (ed.), pp. 59–70. San Francisco: Freeman 1974 [73]

Jung, R.: Correlation of bioelectrical and autonomic phenomena with alterations of consciousness and arousal in man. In: Brain mechanisms and consciousness. Delafresnaye, J.F. (ed.), pp. 310–344. Oxford: Blackwell 1954 [53]

Jung, R.: Hirnpotentialwellen, Neuronenentladungen und Gleichspannungsphänomene. In: Jenenser EEG-Symposion „30 Jahre Elektroenzenphalographie". Werner, R. (ed.), pp. 54–81. Berlin: Volk und Gesundheit 1963 [80]

Jung, R.: Visual perception and neurophysiology. In: Handbook of sensory physiology.

Jung, R. (ed.), Vol. II/3A, pp. 1–152. Berlin, Heidelberg, New York: Springer 1973 [62]
Jung, R.: Perception, consciousness and visual attention. In: Cerebral correlates of conscious experience. Buser, P., Rougeul-Buser, A. (eds.), pp. 15–36. Amsterdam: Elsevier 1978 [53, 69]
Jung, R.: Perception and Action. In: Proceedings of the 28th International Congress of Physiology, Budapest. Part 10, Regulating Functions of the Central Nervous System (1980) [53, 98, 100]
Kerr, F.W.L., Wilson, P.R.: Pain. Annu. Rev. Neurosci. *1*, 83–102 (1978) [128]
Kety, S.S.: Sleep and the energy metabolism of the brain. In: The nature of sleep. Ciba Foundation Symposium. Wolstenholme, G.E.W., O'Connor, M. (eds.), pp. 375–385. London: Churchill 1961 [135]
Kety, S.S.: New perspectives in psychopharmacology. In: Beyond Reductionism. Koestler, A.A., Smythies, J.R. (eds.), pp. 334–347. London: Hutchinson (1969) [26]
Kety, S.S.: The biogenic amines in the central nervous system: their possible roles in arousal, emotion, and learning. In: The neurosciences. Schmitt, F.O. (ed.), pp. 324–336. New York: Rockefeller University Press 1970 [117]
Kety, S.S.: Norepinephrine in the central nervous system and its correlations with behaviour. In: Brain and behaviour. Karczmar, A.G., Eccles, J.C. (eds.), pp. 115–128. Berlin, Heidelberg, New York: Springer 1972 [117]
Kleitman, N.: Patterns of dreaming. In: Frontiers of psychological research. Coopersmith, S. (ed.), pp. 236–242. San Francisco: Freeman 1960 [155]
Kleitman, N.: The nature of dreaming. In: The nature of sleep. Wolstenholme, G.E.W., O'Connor, M. (eds.), pp. 349–364. London: Churchill 1961 [155]
Kleitman, N.: Sleep and wakefulness. Chicago: University of Chicago Press 1963 [155]
Kneale, W.: On having a mind. London: Cambridge University Press 1962 [236]
Kornhuber, H.H.: Neural control of input into long term memory: limbic system and amnestic syndrome in man. In: Memory and transfer of information. Zippel, H.P. (ed.), pp. 1–22. New York: Plenum 1973 [176, 177]
Kornhuber, H.H.: Cerebral cortex, cerebellum, and basal ganglia: An introduction to their motor functions. In: The neurosciences. Third Study Program. Schmitt, F.O., Worden, F.G. (eds.), pp. 267–280. Cambridge, Mass.: MIT Press 1974 [46, 102]
Kornhuber, H.H.: A reconsideration of the brain–mind problem. In: Cerebral correlates of conscious experience. Buser, P.A., Rougeul-Buser, A. (eds.), pp. 319–331. Amsterdam: North Holland 1978 [48]
Kornhuber, H.H., Deecke, L.: Hirnpotentialänderungen beim Menschen vor und nach Willkürbewegungen, dargestellt mit Magnetbandspeicherung und Rückwärtsanalyse. Pflügers Arch. Ges. Physiol. *281*, 52 (1964) [99]
Krnjević, K.: Chemical nature of synaptic transmission in vertebrates. Physiol. Rev. *54*, 418–540 (1974) [146]
Langer, S.K.: Philosophy in a new key. Cambridge, Mass.: Harvard University Press 1951 [217]
Lassen, N.A., Ingvar, D.H., Skinhøj, E.: Patterns of activity in the human cerebral cortex graphically displayed. Sci. Am. *239/4*, 50–59 (1978) [87, 99, 131, 135, 136]
Laszlo, E.: Introduction to systems philosophy. New York: Gordon & Breach 1972 [18]
Lemberg, R.: Complementarity of religion and science. A trialogue. Zygon 1980 [250]

Levy-Agresti, J., Sperry, R.W.: Differential perceptual capacities in major and minor hemisphere. Proc. Natl. Acad. Sci. USA *61*, 1151 (1968) [10]

Lewis, H.D.: Persons and life after death. London: MacMillan 1978 [234, 249]

Libet, B.: Electrical stimulation of cortex in human subjects, and conscious sensory aspects. In: Handbook of sensory physiology. Iggo, A. (ed.), Vol. II, pp. 743–790. Berlin, Heidelberg, New York: Springer 1973 [46, 86, 87, 96, 129, 232]

Libet, B., Wright, E.W., Feinstein, B.: Subjective referral of the timing for a conscious experience: a functional role for the somatosensory specific projection system in man. Brain (1979) [129]

Lorenz, K.: On aggression. London: Methuen *102*, 193–224, 1966 [208]

Lorenz, K.: Behind the mirror. London: Methuen 1977 [26]

Lucas, J.R.: The freedom of the will. London: Oxford University Press 1970 [241, 242]

MacKay, D.M.: Selves and brains. Neurosciences *3*, 599–606 (1978) [14, 24, 25, 106]

MacLean, P.D.: The paranoid streak in man. In: Beyond reductionism. Koestler, A., Smythies, J.R. (eds.), pp. 258–278. London: Hutchinson 1969 [116]

MacLean, P.D.: The triune brain, emotion, and scientific bias. In: The neurosciences: Second study program. Schmitt, F.O. (ed.), pp. 336–349. New York: Rockefeller University Press 1970 [109, 110, 116, 117]

McGeer, P.L., Eccles, J.C., McGeer, E.R.: Molecular neurobiology of the mammalian brain. New York: Plenum 1978 [105, 111, 117, 118, 146]

Magoun, H.W.: The waking brain. Springfield, Ill.: Thomas 1958 [144]

Mandler, G.: An ancient conundrum. Science *200*, 1040–1041 (1978) [242]

Marin-Padilla, M.: Prenatal and early postnatal ontogenesis of the human motor cortex: A Golgi study. II. The basket-pyramidal system. Brain Res. *23*, 185–191 (1970) [39]

Mark, V.H., Ervin, F.R.: Violence and the brain. New York: Harper & Row 1970 [116]

Marr, D.: A theory of the cerebellar cortex. J. Physiol. (London) *202*, 437–470 (1969) [176]

Marr, D.: A theory for cerebral neocortex. Proc. R. Soc. Lond. B *176*, 161–234 (1970) [176]

Marr, D., Poggio, T.: A computational theory of human stereo vision. Proc. R. Soc. Lond. B *204*, 301–328 (1979) [74]

Matsumoto, H., Ayala, G.F., Gumnit, R.J.: Neuronal behaviour and triggering mechanism in cortical epileptic focus. J. Neurophysiol. *32*, 688–703 (1969) [159]

Medawar, P.B.: Induction and intuition in scientific thought. London: Methuen 1969 [188, 189]

Melzack, R., Stotler, W.A., Livingston, W.K.: Effects of discrete brainstem lesions in cats on perception of noxious stimulation. J. Neurophysiol. *21*, 353–367 (1958) [127]

Merzenich, M.M., Brugge, J.F.: Representation of the cochlear partition on the superior temporal plane of the Macaque monkey. Brain Res. *50*, 275–296 (1973) [31]

Merzenich, M.M., Knight, P.L., Roth, G.L.: Representation of the cochlea within primary auditory cortex in the cat. J. Neurophysiol. *38*, 231–249 (1975) [31]

Milner, B.: Hemispheric specialization: scope and limits. In: The Neurosciences: Third study program. Schmitt, F.O., Worden, F.G. (eds.), pp. 75–89. Cambridge, Mass., London: MIT Press 1974 [186]

Misgeld, U., Sarvey, J.M., Klee, M.R.: Heterosynaptic post-activation potentiation in hippocampal CA3 neurons: Long term changes of the postsynaptic potentials. Exp. Brain Res. *37*, 217–229 (1979) [177]

Monod, J.: Chance and necessity. New York: Knoff 1971 [191, 221, 222, 223, 227]

Monod, J.: On chance and necessity. In: Studies in the philosophy of biology. Ayala, F.J., Dobzhansky, T. (eds.), pp. 357–375. London: MacMillan 1974 [224]

Monod, J.: The analysis of scientific method and the logic of scientific discovery. In: The creative process in science and medicine. Krebs, H.A., Shelley, J.H. (eds.), pp. 3–7, 52. Amsterdam: Excerpta Medica 1975 [189]

Montagu, A.: The nature of human aggression. Oxford: Oxford University Press 1976 [208, 209]

Morris, D.: The naked ape. New York: McGraw-Hill 1967 [208]

Morris, D.: The human zoo. New York: McGraw-Hill 1969 [208]

Moruzzi, G.: The functional significance of sleep with particular regard to the brain mechanisms underlying consciousness. In: Brain and conscious experience. Eccles, J.C. (ed.), pp. 345–388. Berlin, Heidelberg, New York: Springer 1966a [164]

Moruzzi, G.: Brain plasticity. In: Brain and conscious experience. Eccles, J.C. (ed.), pp. 555–560. Berlin, Heidelberg, New York: Springer 1966b [164]

Moruzzi, G., Magoun, H.W.: Brain stem reticular formation and activation of the EEG. Electroencephalogr. Clin. Neurophysiol. *1*, 455–473 (1949) [142, 145]

Mountcastle, V.B.: Modality and topographic properties of single neurones of cat's somatic sensory cortex. J. Neurophysiol. *20*, 408–434 (1957) [31]

Mountcastle, V.B.: Medical physiology, Vol. 2, 12th ed. St. Louis: Mosby 1968 [127]

Mountcastle, V.B.: Some neural mechanisms for directed attention. In: Cerebral correlates of conscious experience. Buser, P.A., Rougeul-Buser, A. (eds.), pp. 37–51. Amsterdam: North Holland 1978a [88, 98]

Mountcastle, V.B.: An organizing principle for cerebral function: the unit module and the distributed system. In: The mindful brain, pp. 7–50. Cambridge, Mass. London: MIT Press 1978b [17, 25, 33, 41, 48]

Mountcastle, V.B., Powell, T.P.S.: Neural mechanisms subserving cutaneous sensibility, with special reference to the role of afferent inhibition in sensory perception and discrimination. Bull. Johns Hopkins Hosp. *105*, 201–232 (1959) [125]

Mountcastle, V.B., Lynch, J.C., Georgopoulos, A., Sakata, H., Acuna, C.: Posterior parietal association cortex of the monkey: command functions for operation within extrapersonal space. J. Neurophysiol. *38*, 871–908 (1975) 54, 56, 57, 98, 129]

Nagel, T.: Brain bisection and the unity of consciousness. Synthese *22*, 396–413 (1971) [15]

Nakahama, H.: Pain mechanisms in the central nervous system. Neurophysiological Basis of Anaesthesia *13*, 109–148 (1975) [121]

Newman, J.D.: Single unit analysis of prefrontal cortex in primates. Exp. Brain Res. [Suppl.] *1*, 459–462 (1976) [39]

Noda, H., Adey, W.R.: Firing of neuron pairs in cat association cortex during sleep and wakefulness. J. Neurophysiol. *33*, 672–684 (1970) [151]

Noda, H., Adey, W.R.: Neuron activity in the association cortex of the cat during sleep, wakefulness and anesthesia. Brain Res. *54*, 243–259 (1973) [151, 160]

Olds, J., Milner, P.: Positive reinforcement produced by electrical stimulation of septal area and other regions of rat brain. J. Comp. Physiol. Psychol. *47*, 419–427 (1954)[114]

Passmore, J.: Discussion. In: Connaissance Scientifique et Philosophie. Publications du Deuxienne Centenaire No 4, pp. 286–287. Bruxelles: Palais des Académies 1975 [173]
Peacocke, A.R.: Science and the Christian experiment. London: Oxford University Press 1971 [244]
Pelikan, J.: The shape of death: life, death and immortality in the early fathers. London: MacMillan 1962 [249]
Penfield, W.: The mystery of the mind. Princeton, N.J.: Princeton University Press 1975 [3, 26, 252]
Penfield, W., Boldrey, E.: Somatic motor and sensory representation in the cerebral cortex of man studied by electrical stimulation. Brain *60*, 389–443 (1937) [125]
Penfield, W., Jasper, H.: Epilepsy and the functional anatomy of the human brain. Boston: Little, Brown & Co. 1954 [81, 82, 143, 148, 157, 158]
Pert, C.B., Snyder, S.H.: Properties of opiate receptor binding in rat brain. Proc. Natl. Acad. Sci. USA *70*, 2243–2247 (1973) [128]
Phillips, C.G., Porter, R.: Corticospinal neurones. London: Academic Press 1977 [31]
Phillis, J.W.: The pharmacology of synapses. Oxford: Pergamon 1970 [145, 146]
Planck, M.: The universe in the light of modern physics. London: Allen & Unwin 1937 [223]
Planck, M.: Scientific autobiography and other papers. London: Williams & Norgate 1950 [218, 223, 224, 246, 247]
Plato: The last days of Socrates, Phaedo. London: Penguin 1954 [248, 249]
Plum, F., Duffy, T.E.: The couple between cerebral metabolism and blood flow during seizures. In: Brain work. Ingvar, D.H., Lassen, N.A. (eds.), pp. 196–214. Copenhagen: Munksgaard 1975 [136]
Poggio, G.F., Fisher, B.: Binocular interaction and depth sensitivity of striate and prestriate cortical neurons of the behaving rhesus monkey. J. Neurophysiol. *40*, 1392–1405 (1977) [74]
Poggio, G.F., Mountcastle, V.B.: A study of the functional contributions of the lemniscal and spinothalamic systems to somatic sensibility. Bull. Johns Hopkins Hosp. *106*, 266–316 (1960) [123]
Polanyi, M., Prosch, H.: Meaning. Chicago: University of Chicago Press 1975 [223, 228]
Popper, K.R.: The logic of scientific discovery. London: Hutchinson 1959 [225]
Popper, K.R.: Epistemology without a knowing subject. In: Logic, methodology and philosophy of science. III, ed. by B. van Rootselaar and Staal, J.F. Amsterdam: North Holland 1968a [166, 167, 169, 175]
Popper, K.R.: On the theory of the objective mind. In: Akten des XIV. Internationalen Kongresses für Philosophie. Vol. 1, pp. 25–53. Vienna: Ed. Leo Gabriel, Verlag Herder 1968b [166, 167, 168, 169]
Popper, K.R.: Objective knowledge: An evolutionary approach. Oxford: Clarendon 1972 [21, 166, 217, 218, 224, 241]
Popper, K.R.: The analysis of scientific method and the logic of scientific discovery. Discussion. In: The creative process in science and medicine. Krebs, H.A., Shelley, J.H. (eds.), pp. 18–19. Amsterdam: Excerpta Medica 1975 [189, 190, 218, 224]
Popper, K.R., Eccles, J.C.: The self and its brain. Berlin, Heidelberg, New York:

Springer 1977 [2, 4, 5, 6, 10, 13, 16, 17, 19, 21, 25, 26, 46, 49, 59, 61, 70, 97, 106, 166, 168, 169, 172, 175, 183, 184, 195, 231, 239, 241, 242]

Power, E.: Medieval people. Harmondworth, Middlesex: Penguin 1951 [210]

Pribram, K.H.: Languages of the brain. Englewood Cliffs: Prentice-Hall 1971 [17, 48]

Puccetti, R.: Brain bisection and personal identity. Br. J. Philos. Sci. *24*, 339–355 (1973) [13]

Pugh, G.E.: The biological origin of human values. London: Routledge & Kegan Paul 1978 [219, 220]

Raichle, M.: Sensori-motor area increase of oxygen uptake and blood flow in the human brain during contralateral hand exercise; preliminary observations by the O-15 method. In: Brain work. Ingvar, D.H., Lassen, N.A. (eds.), pp. 372–376. Copenhagen: Munksgaard 1975 [135, 140]

Ratliff, F.: Contour and contrast. Sci. Am. *226/6*, 91–101 (1972) [62]

Reivich, M., Sokoloff, L., Kennedy, C., Des Rosiers, M.: An autoradiographic method for the measurement of local glucose metabolism in the brain. In: Brain work. Ingvar, D.H., Lassen, N.A. (eds.), pp. 377–384. Copenhagen: Munksgaard 1975 [135]

Rensch, B.: Biophilosophy. New York: Columbia University Press 1971 [17, 18, 20, 25, 88]

Rensch, B.: Polynomistic determination of biological processes. In: Studies in the philosophy of biology. Ayala, F.J., Dobzhansky, T. (eds.), pp. 241–258. London: MacMillan 1974 [17, 20, 25]

Réthelyi, M., Szentágothai, J.: Distribution and connections of afferent fibres in the spinal cord. In: Somatosensory system. Iggo, A. (ed.), pp. 207–252. Berlin, Heidelberg, New York: Springer 1973 [123]

Rock, I.: The perception of disorientated figures. In: Image, object, and illusion. Held, R. (ed.), pp. 71–78. San Francisco: Freeman 1974 [68]

Roth, B.: Narcolepsy and hypersomnia: review and classification of 642 personally observed cases. Schweiz. Arch. Neurol. Neurochir. Psychiatr. *119*, 31–41 (1976) [154]

Roth, B.: Narcolepsy and hypersomnia. In: Sleep disorders: diagnosis and treatment. Williams, Karacon, (eds.), pp. 29–59. New York: Wiley & Sons 1978 [155]

de Rougement, D.: Love in the Western world. New York: Harper & Row 1972 [131]

Ryle, G.: The concept of mind. London: Hutchinson's University Library 1949 [3]

Sagan, C.: The dragons of Eden. New York: Ballantine 1977 [210]

Schlag, J.: Reticular influences on the thalamo-cortical activity. In: Handbook of electroencephalography and clinical neurophysiology. Rémond, A., Creutzfeldt, O. (eds.), pp. 119–135. Amsterdam: Elsevier 1974 [103]

Schmidt, R.F.: Control of the access of afferent activity to somatosensory pathways. In: Somatosensory system. Iggo, A. (ed.), pp. 151–206. Berlin, Heidelberg, New York: Springer 1973 [128]

Schrödinger, E.: Mind and matter. London: Cambridge University Press 1958 [223]

Seylaz, J., Pinard, E., Mamo, H., Goas, J.Y., Luft, A., Correze, J.L.: Human cerebral blood flow during sleep. In: Brain work. Ingvar, D.H., Lassen, N.A. (eds.), pp. 235–245. Copenhagen: Munksgaard 1975 [135]

Shapere, D.: Discussion of Rensch. In: Studies in the philosophy of biology. Ayala, F.J., Dobzhansky, T. (eds.), pp. 256–258. London: MacMillan 1974 [20, 25]

Sherrington, C.S.: The brain and its mechanism. Cambridge: Cambridge University Press 1933 [22]
Sherrington, C.S.: Man on his nature. London: Cambridge University Press 1940 [20, 22, 43, 97, 106, 107, 206, 207, 212, 216, 223, 227]
Skolimowski, H.: Problems of rationality in biology. In: Studies in the philosophy of biology. Ayala, F.J., Dobzhansky, T. (eds.), pp. 205–224. London: MacMillan 1974 [223]
Smart, J.J.C.: Philosophy and scientific realism. London: Routledge & Kegan Paul 1963 [18]
Smart, J.J.C.: Presentation at the 16th World Congress of Philosophy, Düsseldorf, 1978 [18, 175]
Solecki, R.S.: Shanidar, New York: Knopf 1971 [204]
Solecki, R.S.: The implications of the Shanidar cave. Neanderthal flower burial. Ann. N.Y. Acad. Sci. *293*, 114–124 (1977) [204]
Spatz, W.B.: Topographically organized reciprocal connections between areas 17 and MT (Visual area of superior temporal sulcus) in the marmoset callithrix jacchus. Exp. Brain Res. *27*, 559–572 (1977) [58]
Sperry, R.W.: Neurology and the mind–brain Problem. Am. Sci. *40*, 291–312 (1952) [22, 234]
Sperry, R.W.: Bridging science and values; a unifying view of mind and brain. In: The centrality of science and absolute values. Proceedings of the fourth international conference on the unity of the sciences. pp. 247–259. New York: International Cultural Foundation 1975 [223, 225]
Sperry, R.W.: Mental phenomena as causal determinants in brain function. In: Consciousness and the brain. Globus, G.G., Maxwell, G., Savodnik, I. (eds.), pp. 163–177. New York: Plenum 1976 [3, 14, 18, 22, 23, 223, 234]
Sperry, R.W.: Forebrain commissurotomy and conscious awareness. J. Med. Phil. *2*, 101–126 (1977) [3, 18, 22]
Sperry, R.W., Zaidel, E., Zaidel, D.: Self-recognition and social awareness in the deconnected minor hemisphere. Neuropsychologia, *17*, 156–166 (1979) [11, 12, 14]
Stark, F.: Perseus and the wind. London: Murray 1948 [250, 251]
Starr, C. (ed.): Biology today. New York: Random House 1975 [197]
Stent, G.: Genes personified. In: The foundations of ethics and its relationship to science. Tristram Engelhardt, H., Callahan, D. (eds.), pp. 33–36. Hastings on Hudson New York: Hastings Center Report 1977 [195]
Stent, G.: Modern Biology and its challenge to philosophy. In: 16th World Congress of Philosophy, Düsseldorf: 1978 [196, 199]
Stent, G.: Does God play dice. Sciences *19/3*, 18–23 (1979) [246]
Storr, A.: Human aggression. New York: Atheneum 1968 [208]
Storr, A.: Human destructiveness. New York: Basic Books 1972 [208]
Strawson, P.: Individuals. London: Methuen 1959 [13]
Sutherland, N.S.: Outlines of a theory of visual pattern recognition in animals and man. Proc. R. Soc. Lond. B *171*, 297–317 (1968) [59]
Szentágothai, J.: The basic neuronal circuit of the neocortex. In: Synchronization of EEG activity in epilepsies. Petsche, H., Brazier, M.A.B. (eds.), pp. 9–24. Vienna, New York: Springer 1972 [34]
Szentágothai, J.: The "module concept" in cerebral cortex architecture. Brain Res. *95*, 475–496 (1975) [34]

Szentágothai, J.: Die Neuronenschaltungen der Großhirnrinde. Verh. Anat. Ges. *70*, 187–215 (1976) [32]

Szentágothai, J.: The neuron network of the cerebral cortex: A functional interpretation. Proc. R. Soc. Lond. B *201*, 219–248 (1978a) [23, 24, 36, 39, 40, 41, 44, 45, 96, 177, 178, 179, 181, 186]

Szentágothai, J.: The local neuronal apparatus of the cerebral cortex. In: Cerebral correlates of conscious experience. Buser, P., Rougeul-Buser, A. (eds.), pp. 131–138. Amsterdam: Elsevier 1978b [23, 37, 44]

Szentágothai, J.: Local neuron circuits of the neocortex. In: The neurosciences: Fourth study program Vol. 4 (1979) [23, 44]

Teuber, M.L.: Sources of ambiguity in the prints of Maurice C. Escher. Sci. Am. *231/1*, 90–104 (1974) [67, 76]

Thach, W.T.: Discharge of cerebellar neurons related to two maintained postures and two prompt movements. I. Nuclear cell output. J. Neurophysiol. *33*, 527–536 (1970) [106]

Thorpe, H.W.: Biology and the nature of man. London: Oxford University Press 1962 [223, 252]

Thorpe, H.W.: Ethology and consciousness. In: Brain and conscious experience. Eccles, J.C. (ed.), pp. 470–505. Berlin, Heidelberg, New York: Springer 1966 [98]

Thorpe, W.H.: Animal nature and human nature. London: Methuen 1974 2, 4, 26, 208, 211, 223, 238]

Thorpe, W.H.: Purpose in a world of chance. Oxford: Oxford University Press 1978 [49, 222, 223, 227, 228]

Tiger, L., Fox, R.: The imperial animal. New York: Rinehart & Winston 1971 [208]

Toyama, K., Matsunami, K., Ohno, T., Tokashiki, S.: An intracellular study of neuronal organization in the visual cortex. Exp. Brain Res. *21*, 45–66 (1974) [44]

Toyama, K., Kimura, M., Shida, T., Takedo, T.: Convergence of retinal inputs onto visual cortical cells. II. A study of the cells disynaptically excited from the lateral geniculate body. Brain Res. *137*, 221–231 (1977a) [44]

Toyama, K., Maekawa, K., Takeda, T.: Convergence of retinal inputs onto visual cortical cells: I. A study of the cells monosynaptically excited from the lateral geniculate body. Brain Res. *137*, 207–220 (1977b) [44]

Uttal, W.R.: The psychobiology of mind. Hillsdale, N. J.: Erlbaum 1978 [25, 196]

Villee, C.A.: Biology. Philadelphia: Saunders 1972 [197]

Vyklický, L., Keller, O., Brožek, G., Butkhuzi, S.M.: Cortical potentials evoked by stimulation of tooth pulp afferents in the cat. Brain Res. *41*, 211–213 (1972) [127]

Wall, P.D.: The gate control theory of pain mechanism. A re-examination and restatement. Brain *101*, 1–18 (1978) [128]

Walter, W.G., Cooper, R., Aldridge, V.J., McCallum, W.D., Winter, A.L.: Contingent negative variation: an electric sign of sensorimotor association and expectancy in the human brain. Nature (Lond.) *203*, 380–384 (1964) [99]

Weiss, P., Discussion. In: Beyond reductionism. Koestler, A., Smythies, J.R. (eds.), pp. 251–252. London: Hutchinson 1969 [5]

Werner, G., Whitsel, B.L.: Functional organization of the somato-sensory cortex. In: Somato-sensory system. Iggo, A. (ed.), pp. 621–700. Berlin, Heidelberg, New York: Springer 1973 [31]

Whitsel, B.L., Petrucelli, L.M., Werner, G.: Symmetry and connectivity in the map of the body surface in somatosensory area II of primates. J. Neurophysiol. *32*, 170–183 (1969) [125]

Whittaker, V.P., Gray, E.G.: The synapse: Biology and morphology. Br. Med. Bull. *18*, 223–228 (1962) [29]

Wigner, E.P.: Two kinds of reality. Monist *48*, 248–264 (1964) [223]

Wigner, E.P.: Are we machines? Proc. Am. Philos. Soc. *113*, 95–101 (1969) [223]

Willis, W.D., Trevino, D.L., Coulter, J.D., Maunz, R.A.: Responses of primate spinothalamic tract neurons to natural stimulation of hindlimb. J. Neurophysiol. *37*, 358–372 (1974) [123]

Wilson, D.L.: On the nature of consciousness and of physical reality. Perspect. Biol. Med. *19*, 568–581 (1976) [17]

Wilson, E.P.: Sociobiology: The new synthesis. Cambridge, Mass.: Belknap, Harvard University Press 1975 [2, 194, 196, 200, 201, 202, 203, 209]

Wilson, E.O.: On human nature. Cambridge, Mass.: Harvard University Press 1978 [194, 196, 205, 212]

Wilson, E.O., Harris, M.: The envelope and the twig. Sciences *18*, 10–15, 27 (1978) [195]

Yarbus, A.L.: Eye movements and vision. New York: Plenum 1967 [59, 60]

Zaidel, E.: Unilateral auditory language comprehension on the token test following cerebral commissurotomy and hemispherectomy. Neuropsychologia *15*, 1–18 (1976) [14]

Zangwill, O.L.: Thought and the brain. Br. J. Psychol. *67*, 301–314 (1976) [3]

Zeki, S.: Functional organization of a visual area in the posterior bank of the superior temporal sulcus of the rhesus monkey. J. Physiol. (London) *236*, 549–573 (1974) [58]

Zeki, S.M.: Projections to superior temporal sulcus from areas 17 and 18 in rhesus monkey. Proc. R. Soc. Lond. B *193*, 199–207 (1976) [58]

Zeki, S.M.: Colour coding in the superior temporal sulcus of rhesus monkey visual cortex. Proc. R. Soc. Lond. B *197*, 195–223 (1977) [58]

Subject Index

Numbers in bold-face type give pages of glossary-type explanations of terms

Experimental Brain Research

ISSN 0014-4819 Title No. 221

Research on the central nervous system has deveoped during the last 20 years into a broad interdisciplinary field. This journal covers the whole field of experimental brain research and thus reflects the variety of interests as presented by the International Brain Research Organization (IBRO): Neurobiology of the Brain.

Fields of Interest: Neurophysiology, Sensory Physiology, Neuroanatomy, Neuropharmacology, Neuropathology, Neurochemistry, Neuroendocrinology, Cellular Biology, Developmental Neurobiology, Neurophysiology and the Behavioral Sciences.

For subscription information and sample copy please write to:
Springer-Verlag, Journal Promotion, Postfach 105 280, D-6900 Heidelberg 1

Springer-Verlag
Berlin Heidelberg New York

MIX
Papier aus verantwortungsvollen Quellen
Paper from responsible sources
FSC® C105338

If you have any concerns about our products,
you can contact us on
ProductSafety@springernature.com

In case Publisher is established outside the EU,
the EU authorized representative is:
Springer Nature Customer Service Center GmbH
Europaplatz 3, 69115 Heidelberg, Germany

Printed by Libri Plureos GmbH
in Hamburg, Germany